1977
Yearbook of Astronomy

1977 Yearbook of Astronomy

Edited by
PATRICK MOORE

Sidgwick & Jackson Limited
LONDON

W. W. Norton & Company Inc
NEW YORK

First Published in Great Britain 1976

Copyright © 1976 by Sidgwick and Jackson Limited

Published in Great Britain by
Sidgwick and Jackson Limited
1 Tavistock Chambers, Bloomsbury Way
London, WC1A 2SG
0.283.98324.8 (cloth)
0.283.98325.X (paper)

First American edition 1977
Published in the United States of America by
W. W. Norton & Company Inc.,
500 Fifth Avenue,
New York 10036

ISBN 0-393-06412-3

Printed in Great Britain
by Eden Fisher (Southend) Ltd

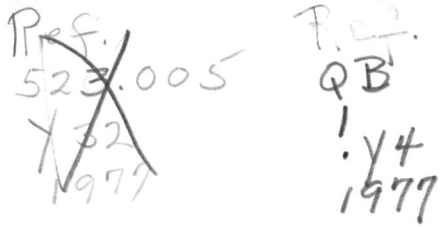

Contents

Editor's Foreword · 7

Preface · 9

PART ONE: EVENTS OF 1977

Notes on the Star Charts for the Northern Hemisphere · 13

The Star Charts (Northern Hemisphere) · 16

Events of 1977 · 42

The Planets in 1977 · 43

The Planets and the Ecliptic · 45

Notes on the Planets in the Monthly Diagrams · 47

Phases of the Moon, 1977 · 50

Monthly Notes:
  Northern Hemisphere · 51
  Southern Hemisphere · 109

Eclipses in 1977 · 126

Occultations in 1977 · 127

Comets in 1977 · 129

Minor Planets in 1977 · 133

Some Events in 1978 · 135

CONTENTS

PART TWO: ARTICLE SECTION

The Rev John Flamsteed – Astronomer Royal and Rector of Burstow: · *E. R. Turner* · 139

The Surface of Venus Revealed: · *H. G. Miles* · 146

Io, the Anomaly of the Solar System: · *G. E. Hunt* · 154

The Unseen Stellar Neighbourhood: · *Peter van de Kamp* · 163

Ice in Space: · *M. Cohen* · 170

The Dusty Sky: H II Regions: · *David A. Allen* · 177

Viking to Mars: · *Patrick Moore* · 183

Recent Advances in Astronomy: · *Patrick Moore* · 193

PART THREE: MISCELLANEOUS

Some Interesting Telescopic Variable Stars · 199

Some Interesting Double Stars · 200

Some Interesting Clusters and Nebulæ · 201

Some Recent Books · 202

Our Contributors · 203

Editor's Foreword

There is one significant change in this, the sixteenth *Yearbook*. Monthly notes, with diagrams, have been added for Southern Hemisphere observers; there have been many requests for this, and it is hoped that the addition will be found useful. Authors of the articles include both our regular contributors and some distinguished newcomers.

As always, Dr J. G. Porter has provided all the essential Northern Hemisphere information for Part 1. My thanks as Editor are due to him, and also to W. D. Procter of Messrs Sidgwick & Jackson, who has been a tower of strength.

PATRICK MOORE

Selsey, 1976

Preface

New readers will find that all the information in this *Yearbook* is given in diagrammatic or descriptive form; the positions of the planets may easily be found on the specially designed star charts, while the monthly notes describe the movements of the planets, and give details of other astronomical phenomena that may be observed from these latitudes. The reader who needs more detailed information will find *Norton's Star Atlas* (Gall and Inglis) an invaluable guide, while more precise positions of the planets and their satellites, together with predictions of occultations, meteor showers and periodic comets may be found in the *Handbook* of the British Astronomical Association. A somewhat similar publication is the *Observer's Handbook* of the Royal Astronomical Society of Canada, and readers will also find details of forthcoming events given in the American *Sky and Telescope*. This monthly publication also produces a special Occultation Supplement giving predictions for the United States and Canada.

Important Note

The star charts are drawn, and the notes are, in general, designed for use in latitude 52 degrees north, but may be used without alteration throughout the British Isles, and (except in the case of eclipses and occultations) in other countries of similar north latitude.

The times given on the star charts and in the Monthly Notes are generally given as local times, using the 24-hour clock, the day beginning at midnight. Ignoring small differences of longitude, this local time may be taken as Greenwich Mean Time (G.M.T.) in the British Isles, or as the appropriate Standard Time in other

PREFACE

Time Zones. If Summer Time is in use, the clocks will have been advanced by one hour, and this hour must be subtracted from the clock time to give G.M.T.

In Great Britain and N. Ireland, Summer Time will be in force in 1977 from 20 March 02^h to 23 October 02^h G.M.T.

The times of a few events (e.g., eclipses) are given in G.M.T., and this may be converted to Local Mean Time by subtracting the west longitude (or adding the east longitude).

    Local Mean Time = G.M.T. − West Longitude
Similarly,
    Eastern Standard Time = G.M.T. − 5 hours,
    Central Standard Time = G.M.T. − 6 hours, etc.

PART ONE

Events of 1977

Monthly Charts and Astronomical Phenomena

Notes on the Star Charts

for the Northern Hemisphere

The stars, together with the Sun, Moon and planets seem to be set on the surface of the celestial sphere, which appears to rotate about the Earth from east to west. Since it is impossible to represent a curved surface accurately on a plane, any kind of star map is bound to contain some form of distortion. But it is well known that the eye can endure some kinds of distortion better than others, and it is particularly true that the eye is most sensitive to deviations from the vertical and horizontal. For this reason the star charts given in this volume on pages 16 to 41 have been designed to give a true representation of vertical and horizontal lines, whatever may be the resulting distortion in the shape of a constellation figure. It will be found that the amount of distortion is, in general, quite small, and is only obvious in the case of large constellations such as Leo and Pegasus, when these appear at the top of the charts, and so are drawn out sideways.

The charts show all stars down to the fourth magnitude, together with a number of fainter stars which are necessary to define the shape of a constellation. There is no standard system for representing the outlines of the constellations, and triangles and other simple figures have been used to give outlines which are easy to follow with the naked eye. The names of the constellations are given, together with the proper names of the brighter stars. The apparent magnitudes of the stars are indicated roughly by using four different sizes of dots, the larger dots representing the bright stars.

There are four such charts at each opening, and these give a complete coverage of the sky up to an altitude of $62\frac{1}{2}$ degrees; there are twelve such sets to cover the entire year. The upper two charts show the southern sky, south being at the centre; the coverage is 200 degrees in azimuth, from a little north of east (top left) to a little north of west (top right). The two lower charts show the northern sky, from a little south of west (lower left) to a little south of east (lower right). There is thus an overlap east and west.

The charts have been drawn for a latitude of 52 degrees, but may be taken without appreciable error to apply to all parts of the British Isles. They will also be equally suitable for any other part of the world having a north latitude of about 52 degrees – e.g. parts of Europe and Asia, and Canada. In such cases the times given must be taken as local time, and not G.M.T., which applies only to the British Isles.

Because the sidereal day is shorter than the solar day, the stars appear to rise and set about four minutes earlier each day, which amounts to two hours in a month. Hence, the twelve sets of charts are sufficient to give the appearance of the sky throughout the day at intervals of two hours, or at the same time of night at monthly intervals throughout the year. The actual range of dates and times when the stars on the charts are visible is indicated at the top of each page. This information is summarized in the following table, which gives the number of the star chart to be used for any given month and time.

G.M.T.	16^h	18^h	20^h	22^h	0^h	2^h	4^h	6^h
January	10	11	12	1	2	3	4	5
February		12	1	2	3	4	5	6
March			2	3	4	5	6	
April			3	4	5	6		
May			4	5	6	7		
June			5	6	7			
July			6	7	8	9		
August			7	8	9	10	11	
September		7	8	9	10	11	12	
October		8	9	10	11	12	1	
November	8	9	10	11	12	1	2	3
December	9	10	11	12	1	2	3	4

NOTES ON THE STAR CHARTS

The charts are drawn to scale, and estimates of altitude and azimuth may be made from them. These values will necessarily be mere approximations, since no observer will be exactly on the meridian of Greenwich at 52 degrees latitude, but they will generally serve for the identification of stars and planets. The horizontal measurements, marked at every ten degrees, give the azimuths (or true bearings) measured from the north round through east (90 degrees), south (180 degrees), and west (270 degrees). The vertical measurements, similarly marked, give the altitudes of the stars up to $62\frac{1}{2}$ degrees.

The ecliptic is drawn as a broken line on which the longitude is marked at every ten degrees; the position of the planets at any time are then easily found by reference to the table on page 43.

There is a curious illusion that stars at an altitude of 60 degrees or more are actually overhead, and the beginner may often feel that he is leaning over backwards in trying to see them. These high-altitude stars, being nearer the pole, move more slowly across the sky, and a different kind of map may therefore be used. These overhead stars are given separately on pages 40 and 41, the entire year being covered at one opening. Each of the four maps shows the overhead stars at times which correspond to those of three of the main star charts. The position of the zenith in latitude 52 degrees is indicated by a cross, and this cross also marks the centre of a circle which is 35 degrees from the zenith, and which therefore indicates an altitude of 55 degrees; there is thus a small overlap with the main charts.

The broken line leading from north to south is numbered to indicate the corresponding main chart. Thus on page 40 the N-S line numbered 6 is to be regarded as an extension of the S line of chart 6 on pages 26 and 27, and at the top of these pages are given dates and times which are appropriate.

The scale is the same on all the charts (approximately 25 degrees to the inch), but the overhead stars are plotted as a true map on a conical projection, and are not simple graphs like the main charts.

1L

October 6 at 5ʰ	October 21 at 4ʰ
November 6 at 3ʰ	November 21 at 2ʰ
December 6 at 1ʰ	December 21 at midnight
January 6 at 23ʰ	January 21 at 22ʰ
February 6 at 21ʰ	February 21 at 20ʰ

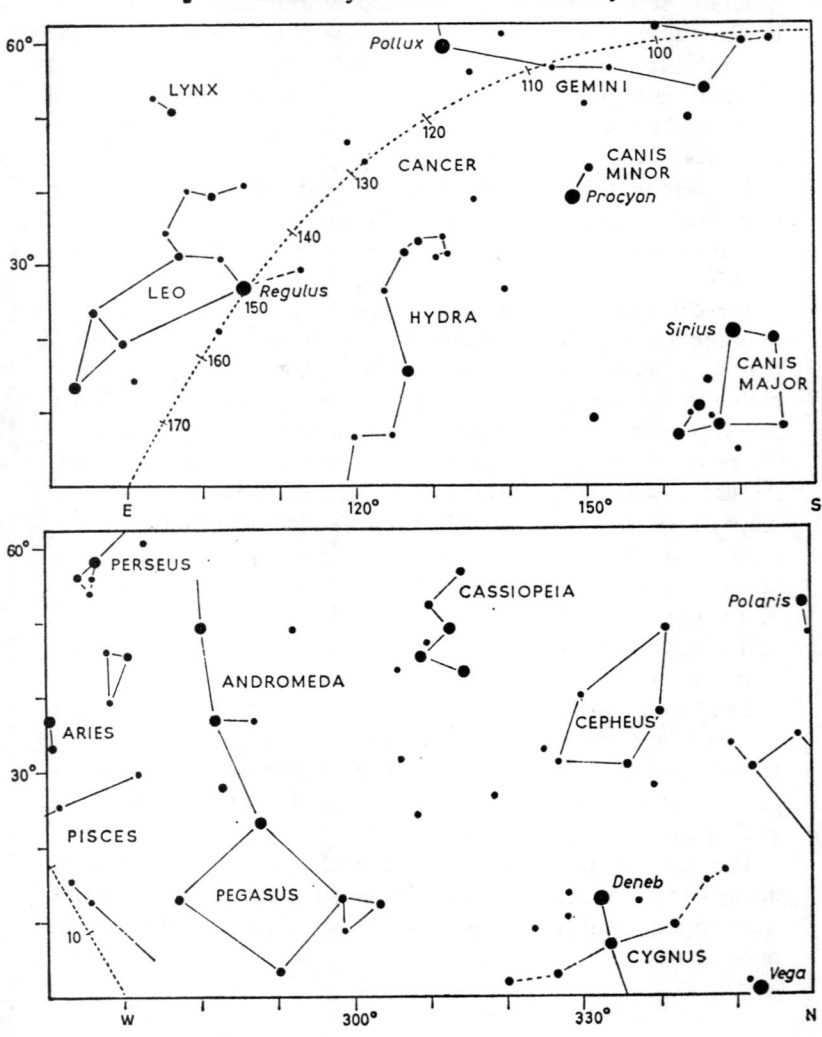

THE STAR CHARTS

October 6 at 5^h	October 21 at 4^h	
November 6 at 3^h	November 21 at 2^h	**1R**
December 6 at 1^h	December 21 at midnight	
January 6 at 23^h	January 21 at 22^h	
February 6 at 21^h	February 21 at 20^h	

2L

November 6 at 5^h	November 21 at 4^h
December 6 at 3^h	December 21 at 2^h
January 6 at 1^h	January 21 at midnight
February 6 at 23^h	February 21 at 22^h
March 6 at 21^h	March 21 at 20^h

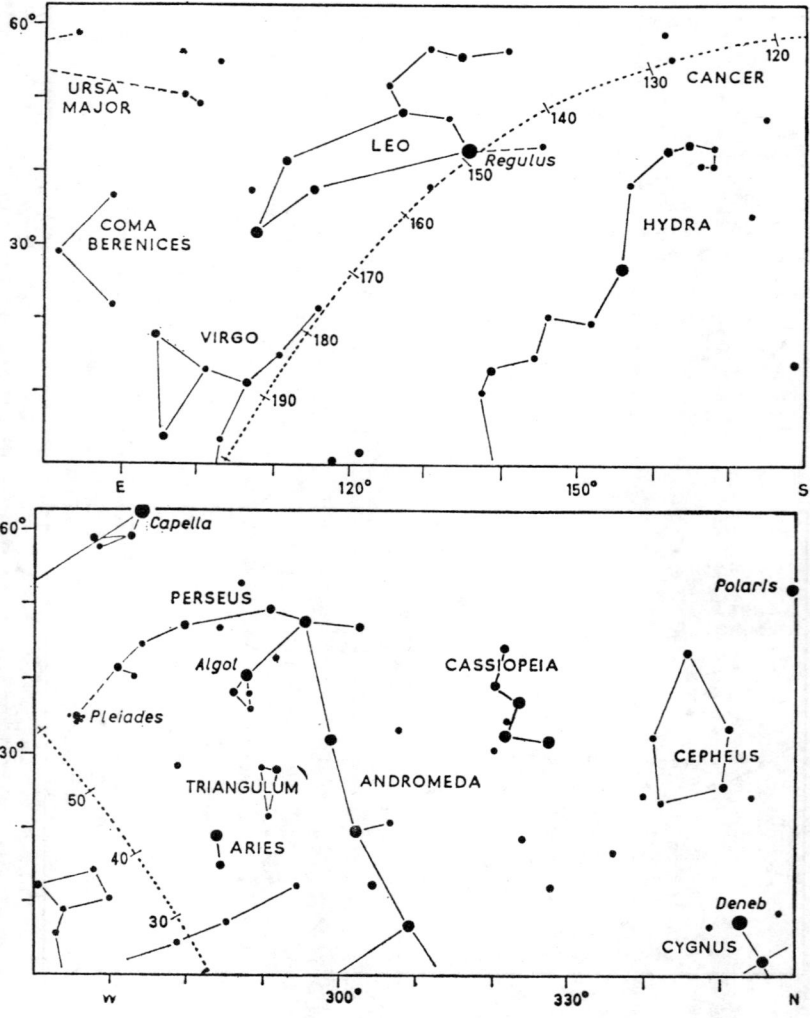

THE STAR CHARTS

November 6 at 5ʰ	November 21 at 4ʰ
December 6 at 3ʰ	December 21 at 2ʰ
January 6 at 1ʰ	January 21 at midnight
February 6 at 23ʰ	February 21 at 22ʰ
March 6 at 21ʰ	March 21 at 20ʰ

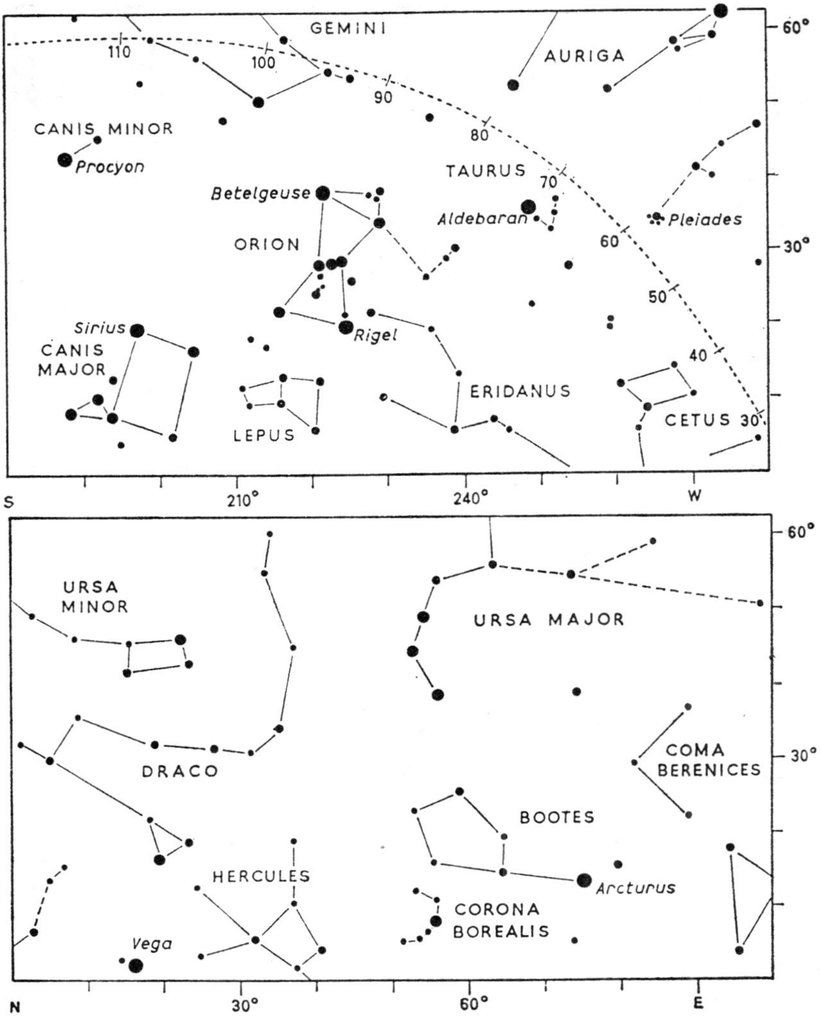

2R

3L

December 6 at 5ʰ	December 21 at 4ʰ
January 6 at 3ʰ	January 21 at 2ʰ
February 6 at 1ʰ	February 21 at midnight
March 6 at 23ʰ	March 21 at 22ʰ
April 6 at 21ʰ	April 21 at 20ʰ

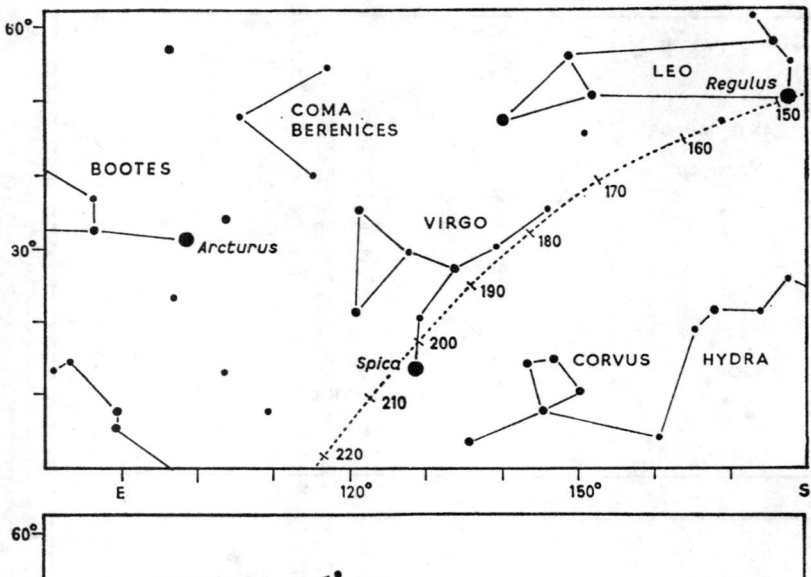

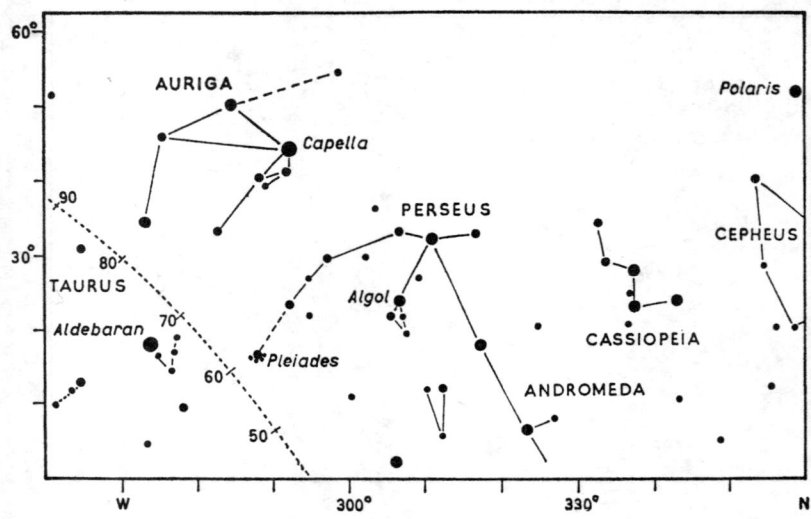

THE STAR CHARTS

December 6 at 5ʰ December 21 at 4ʰ
January 6 at 3ʰ January 21 at 2ʰ
February 6 at 1ʰ February 21 at midnight
March 6 at 23ʰ March 21 at 22ʰ
April 6 at 21ʰ April 21 at 20ʰ

3R

1977 YEARBOOK OF ASTRONOMY

4L

January 6 at 5^h	January 21 at 4^h
February 6 at 3^h	February 21 at 2^h
March 6 at 1^h	March 21 at midnight
April 6 at 23^h	April 21 at 22^h
May 6 at 21^h	May 21 at 20^h

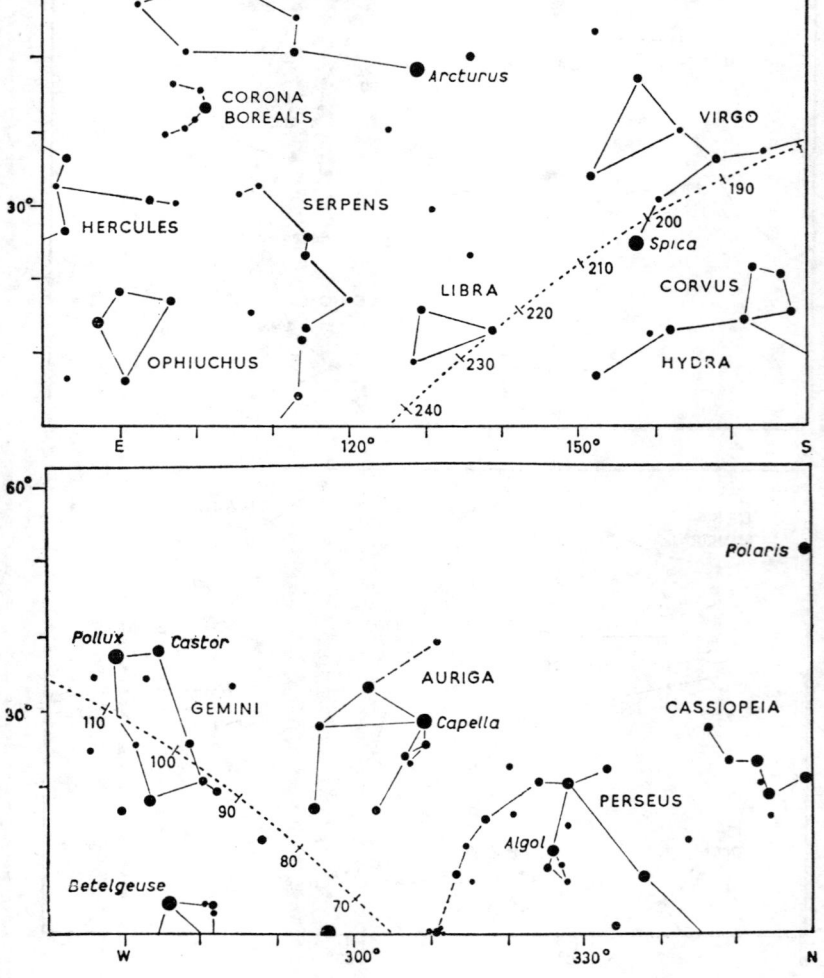

THE STAR CHARTS

January 6 at 5^h	January 21 at 4^h	
February 6 at 3^h	February 21 at 2^h	**4R**
March 6 at 1^h	March 21 at midnight	
April 6 at 23^h	April 21 at 22^h	
May 6 at 21^h	May 21 at 20^h	

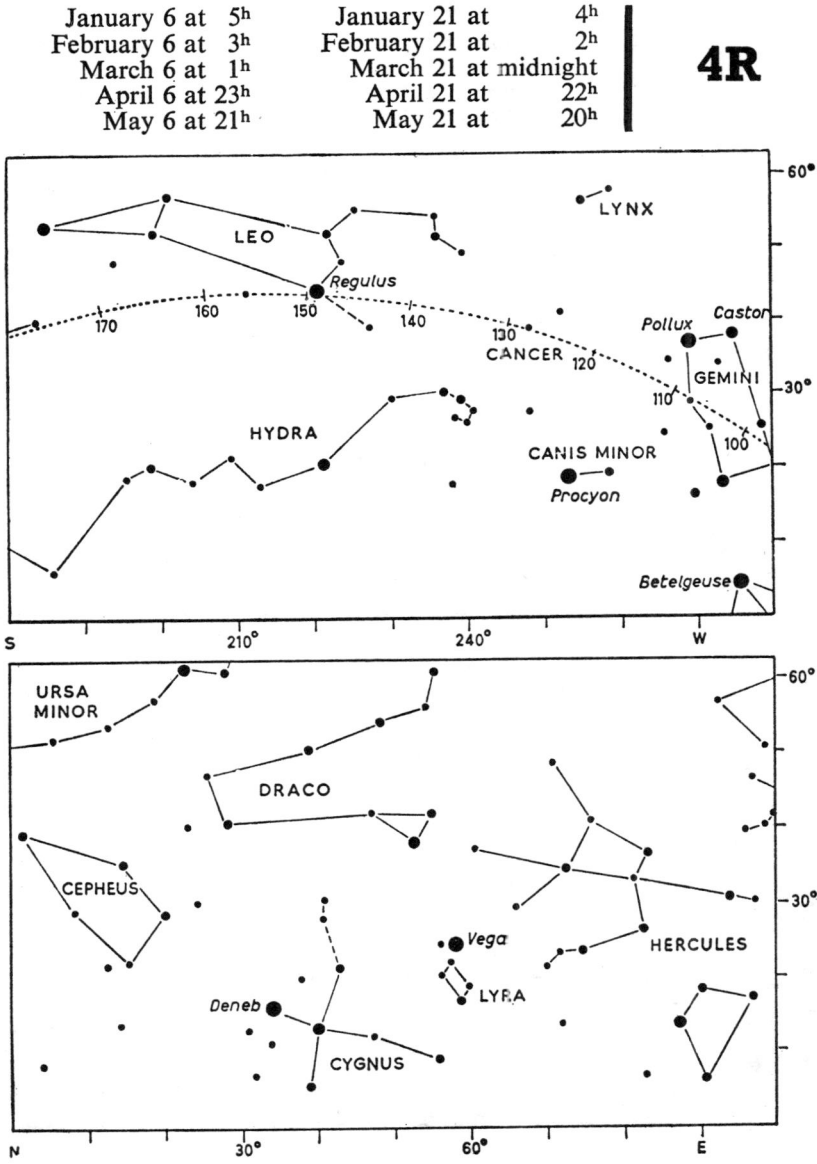

1977 YEARBOOK OF ASTRONOMY

5L

January 6 at 7^h	January 21 at 6^h
February 6 at 5^h	February 21 at 4^h
March 6 at 3^h	March 21 at 2^h
April 6 at 1^h	April 21 at midnight
May 6 at 23^h	May 21 at 22^h

THE STAR CHARTS

January 6 at 7ʰ   January 21 at 6ʰ
February 6 at 5ʰ   February 21 at 4ʰ
March 6 at 3ʰ   March 21 at 2ʰ   **5R**
April 6 at 1ʰ   April 21 at midnight
May 6 at 23ʰ   May 21 at 22ʰ

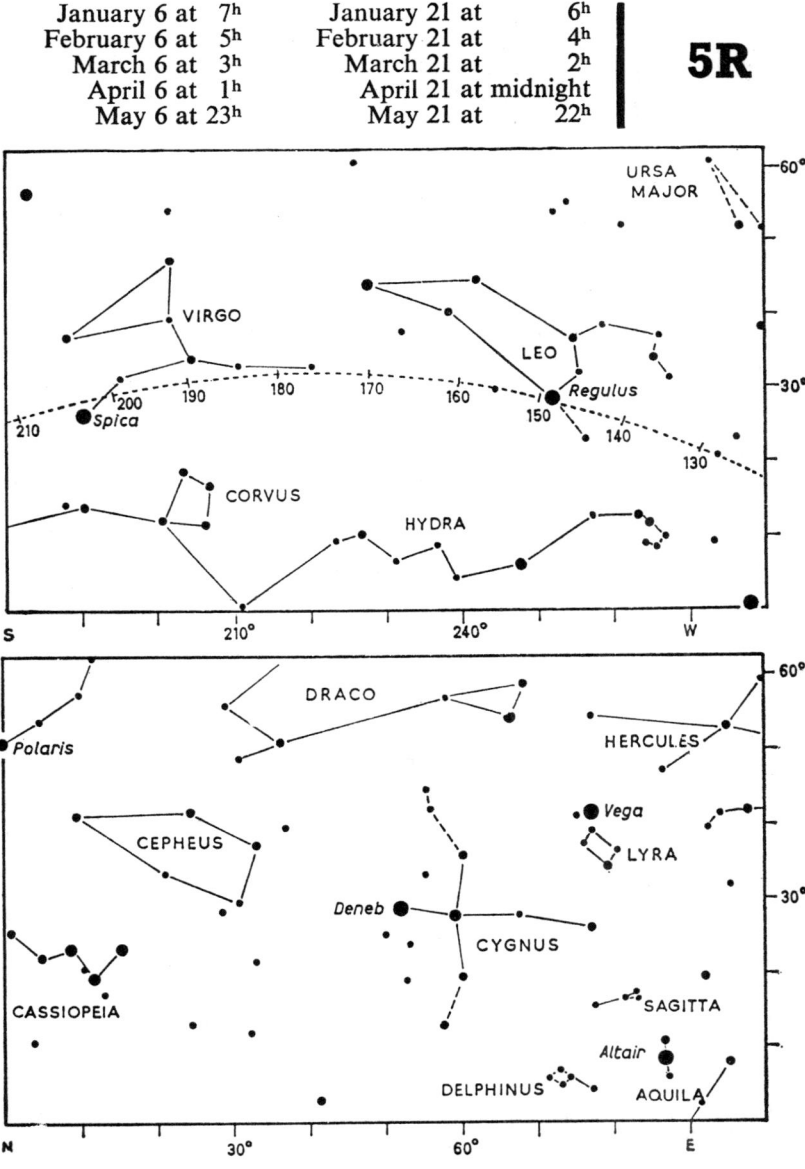

6L

March 6 at	5^h	March 21 at	4^h
April 6 at	3^h	April 21 at	2^h
May 6 at	1^h	May 21 at	midnight
June 6 at	23^h	June 21 at	22^h
July 6 at	21^h	July 21 at	20^h

THE STAR CHARTS

March 6 at 5ʰ	March 21 at 4ʰ	
April 6 at 3ʰ	April 21 at 2ʰ	**6R**
May 6 at 1ʰ	May 21 at midnight	
June 6 at 23ʰ	June 21 at 22ʰ	
July 6 at 21ʰ	July 21 at 20ʰ	

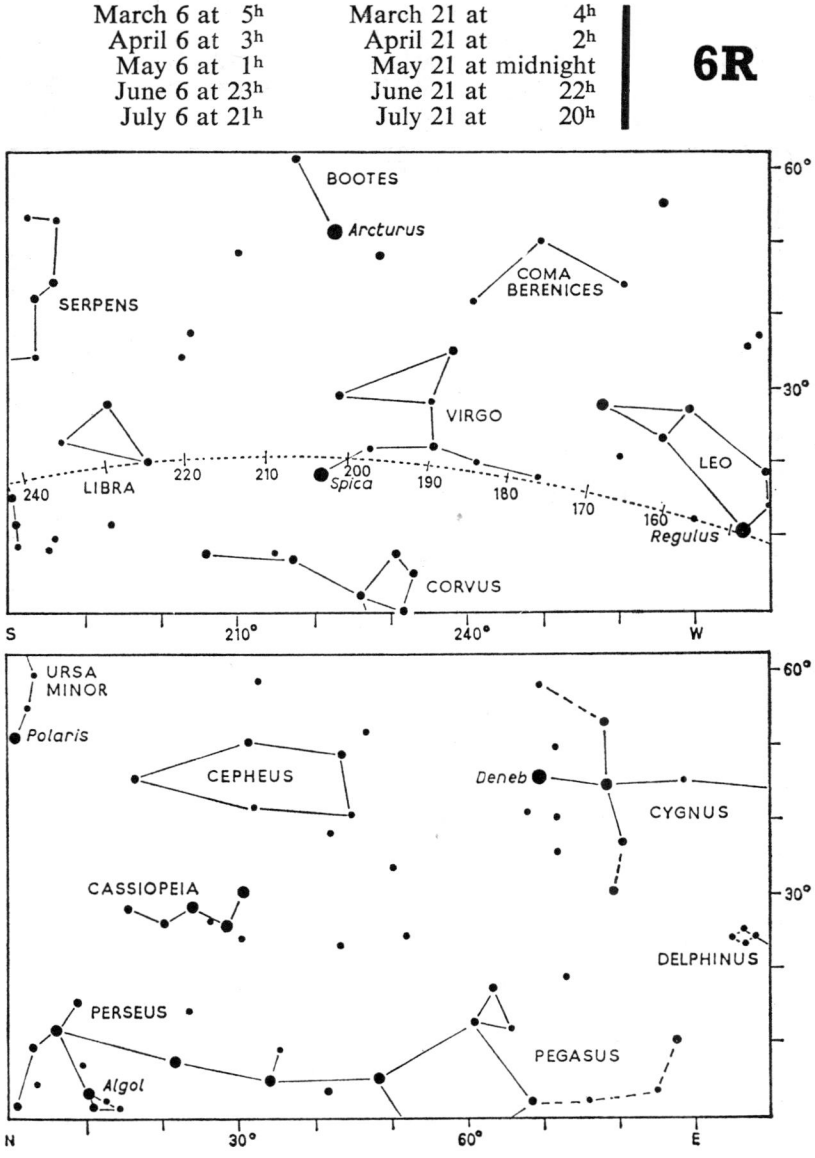

1977 YEARBOOK OF ASTRONOMY

7L

May 6 at 3^h	May 21 at 2^h
June 6 at 1^h	June 21 at midnight
July 6 at 23^h	July 21 at 22^h
August 6 at 21^h	August 21 at 20^h
September 6 at 19^h	September 21 at 18^h

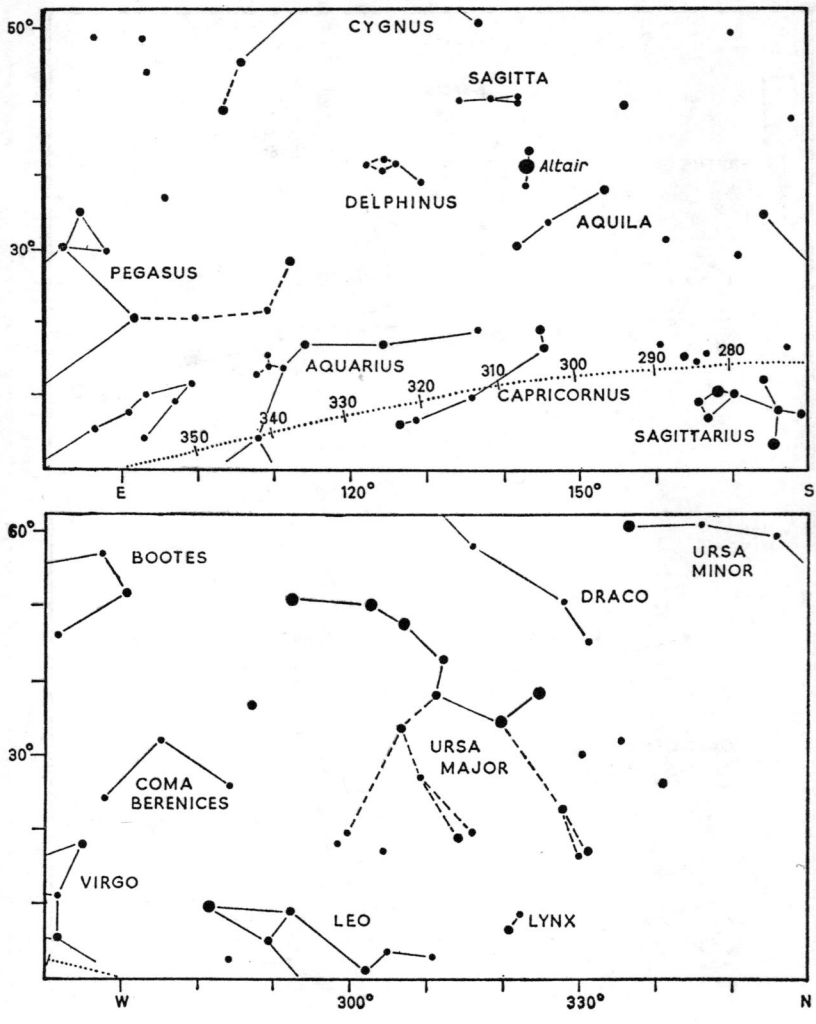

THE STAR CHARTS

May 6 at 3ʰ
June 6 at 1ʰ
July 6 at 23ʰ
August 6 at 21ʰ
September 6 at 19ʰ

May 21 at 2ʰ
June 21 at midnight
July 21 at 22ʰ
August 21 at 20ʰ
September 21 at 18ʰ

7R

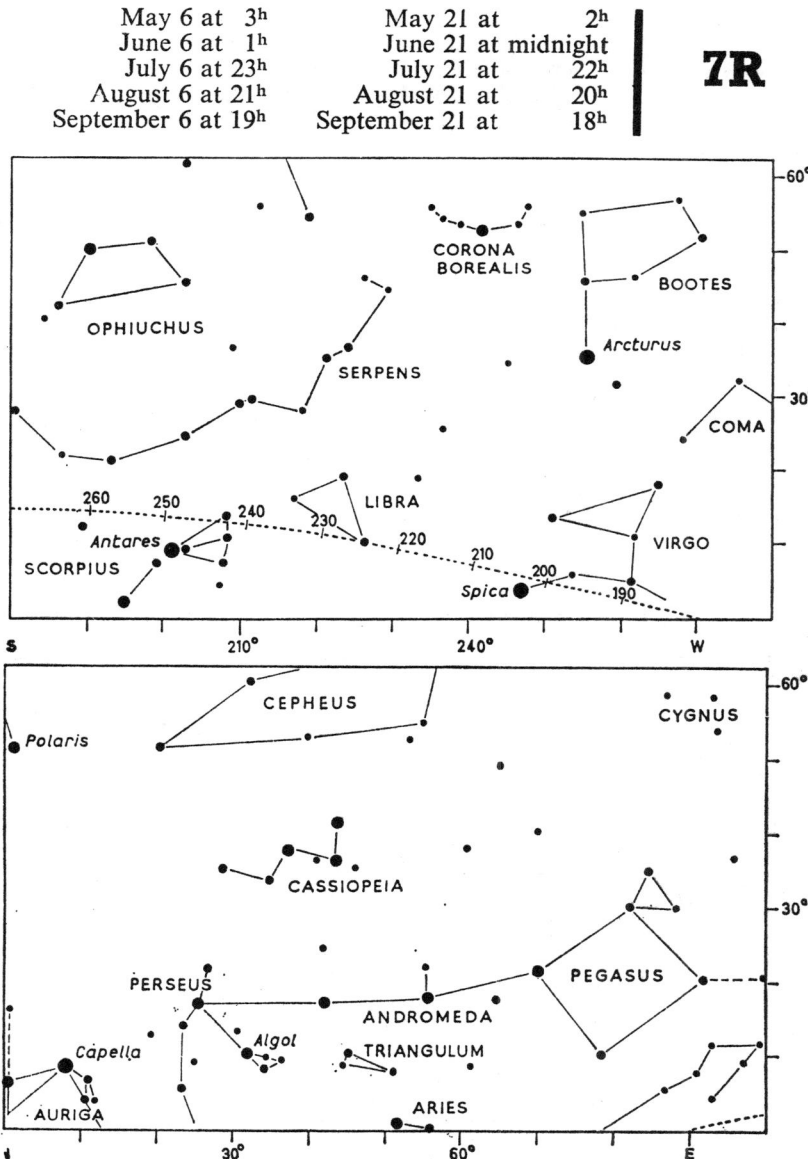

8L

July 6 at 1ʰ	July 21 at midnight
August 6 at 23ʰ	August 21 at 22ʰ
September 6 at 21ʰ	September 21 at 20ʰ
October 6 at 19ʰ	October 21 at 18ʰ
November 6 at 17ʰ	November 21 at 16ʰ

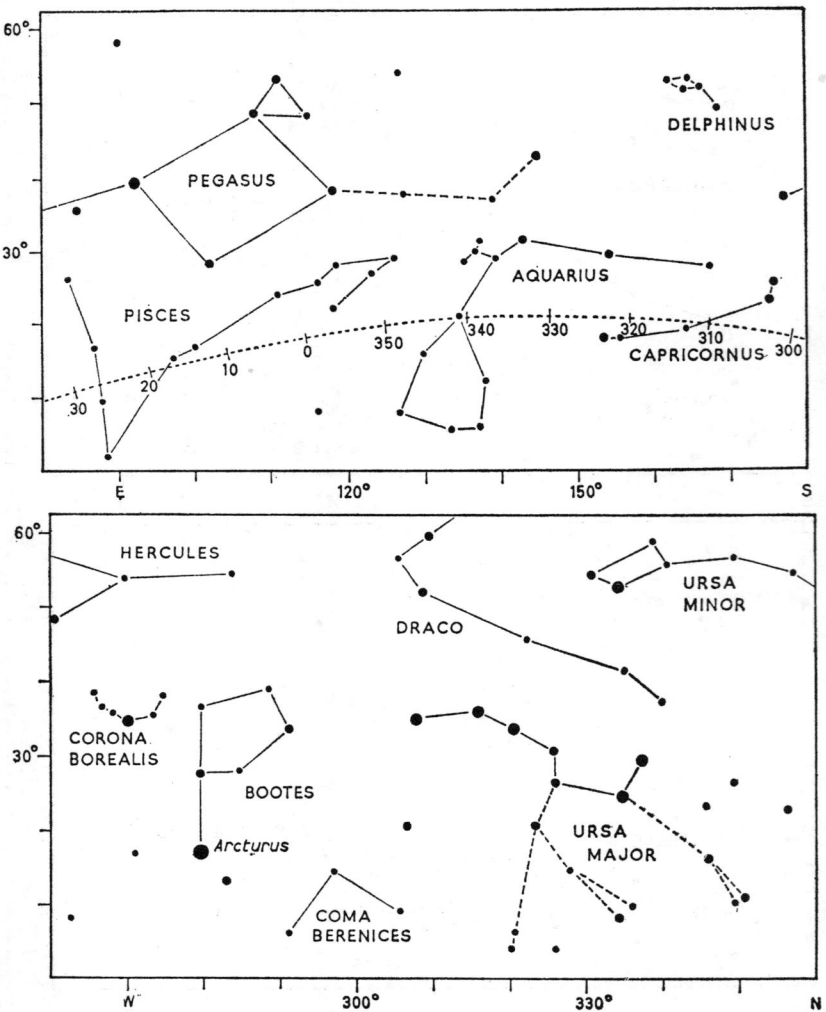

THE STAR CHARTS

July 6 at 1ʰ   July 21 at midnight
August 6 at 23ʰ   August 21 at 22ʰ
September 6 at 21ʰ   September 21 at 20ʰ   **8R**
October 6 at 19ʰ   October 21 at 18ʰ
November 6 at 17ʰ   November 21 at 16ʰ

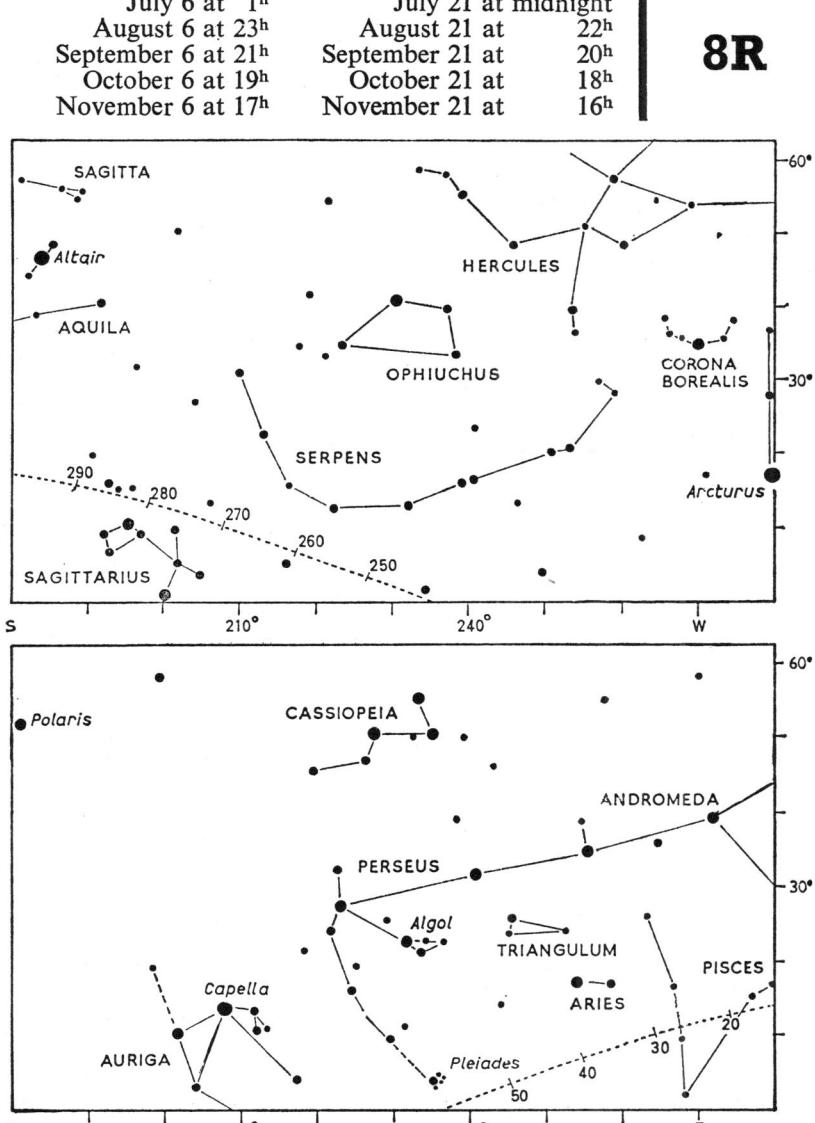

9L

August 6 at 1ʰ	August 21 at midnight
September 6 at 23ʰ	September 21 at 22ʰ
October 6 at 21ʰ	October 21 at 20ʰ
November 6 at 19ʰ	November 21 at 18ʰ
December 6 at 17ʰ	December 21 at 16ʰ

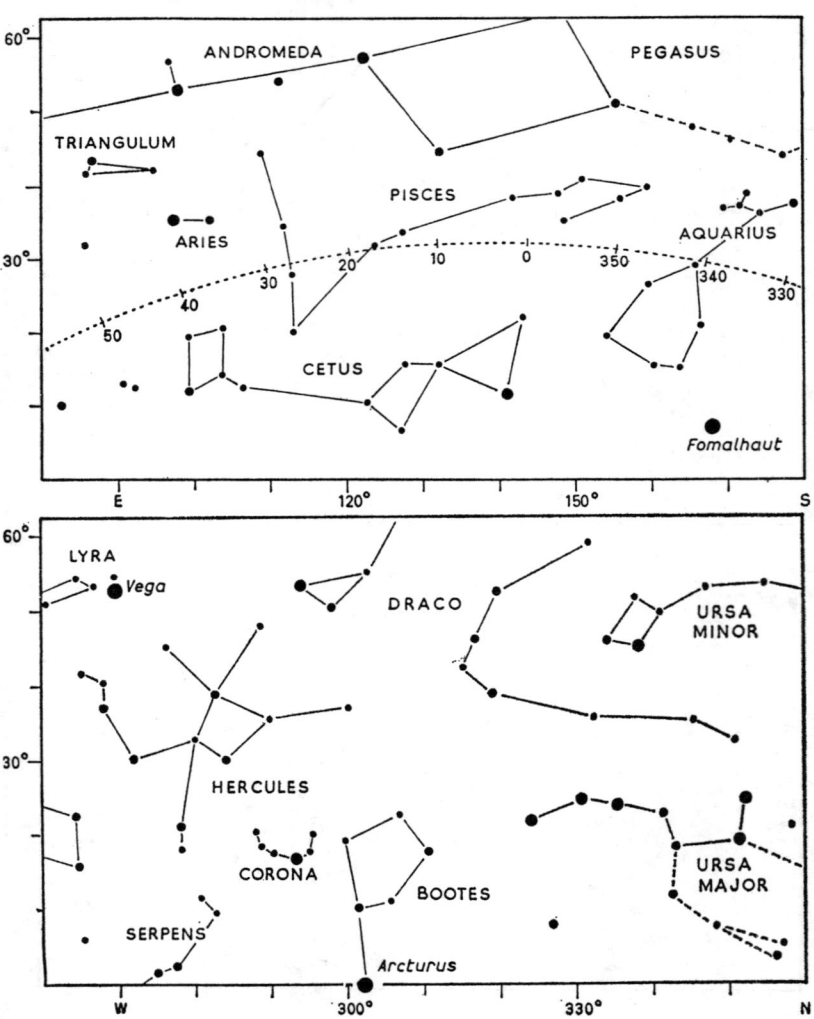

THE STAR CHARTS

August 6 at 1ʰ	August 21 at midnight	
September 6 at 23ʰ	September 21 at 22ʰ	
October 6 at 21ʰ	October 21 at 20ʰ	**9R**
November 6 at 19ʰ	November 21 at 18ʰ	
December 6 at 17ʰ	December 21 at 16ʰ	

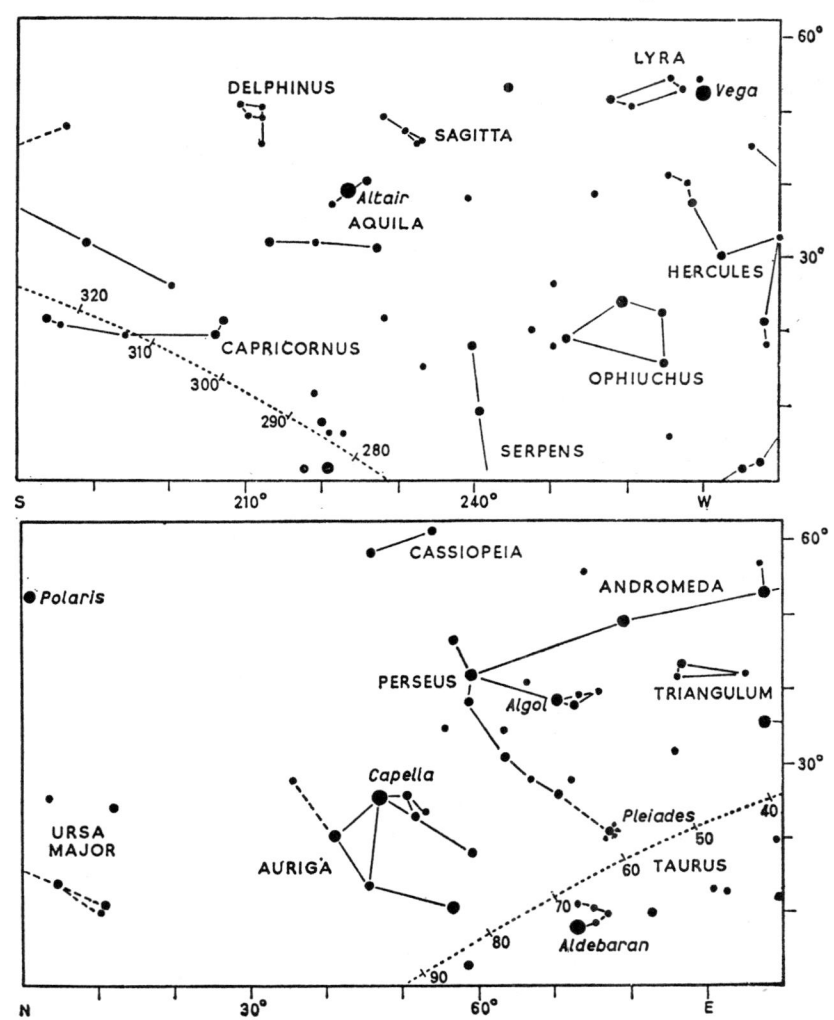

1977 YEARBOOK OF ASTRONOMY

10L

August 6 at 3^h	August 21 at 2^h
September 6 at 1^h	September 21 at midnight
October 6 at 23^h	October 21 at 22^h
November 6 at 21^h	November 21 at 20^h
December 6 at 19^h	December 21 at 18^h

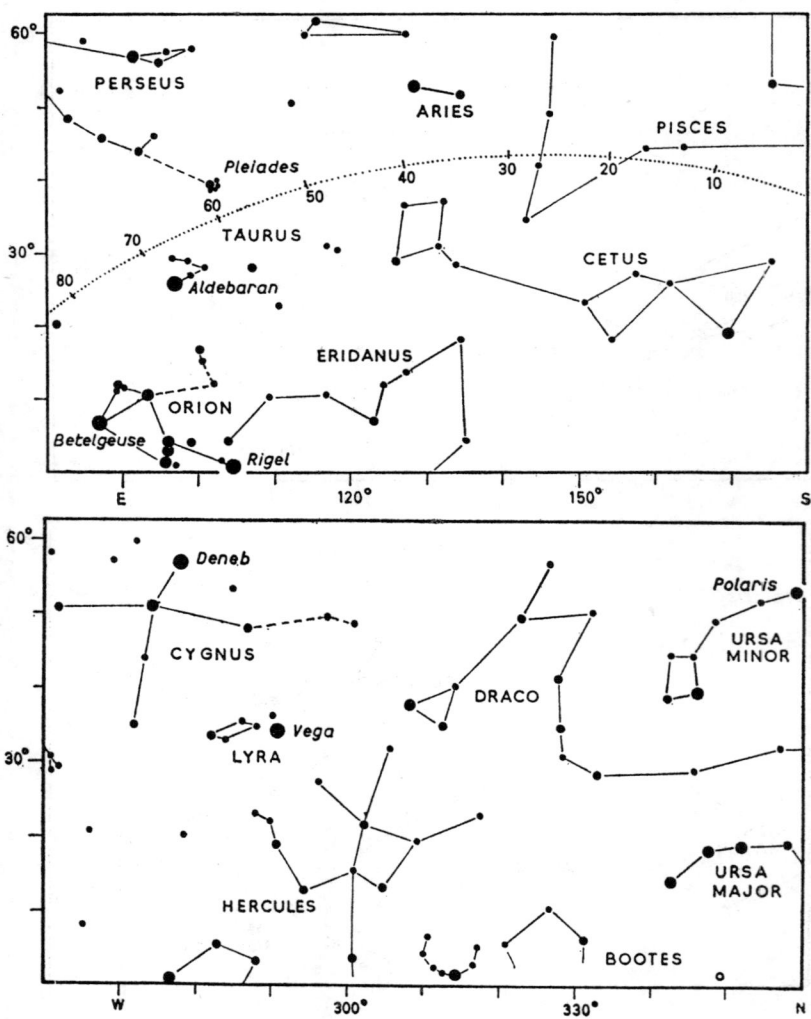

THE STAR CHARTS

August 6 at 3ʰ
September 6 at 1ʰ
October 6 at 23ʰ
November 6 at 21ʰ
December 6 at 19ʰ

August 21 at 2ʰ
September 21 at midnight
October 21 at 22ʰ
November 21 at 20ʰ
December 21 at 18ʰ

10R

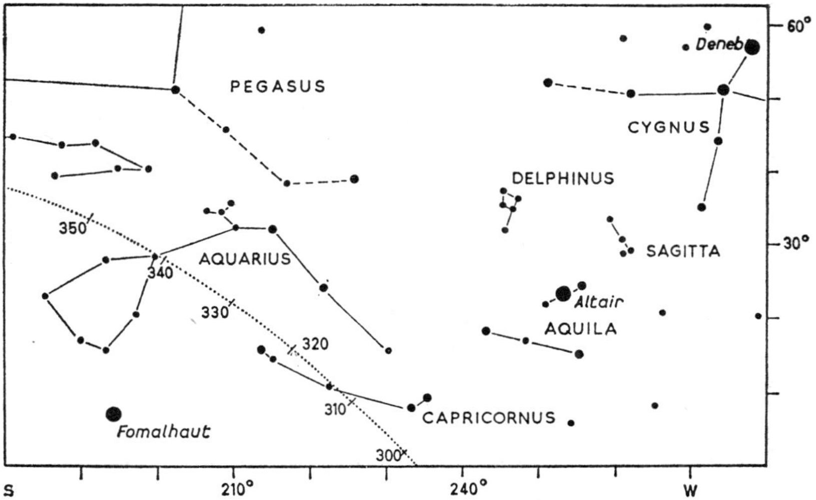

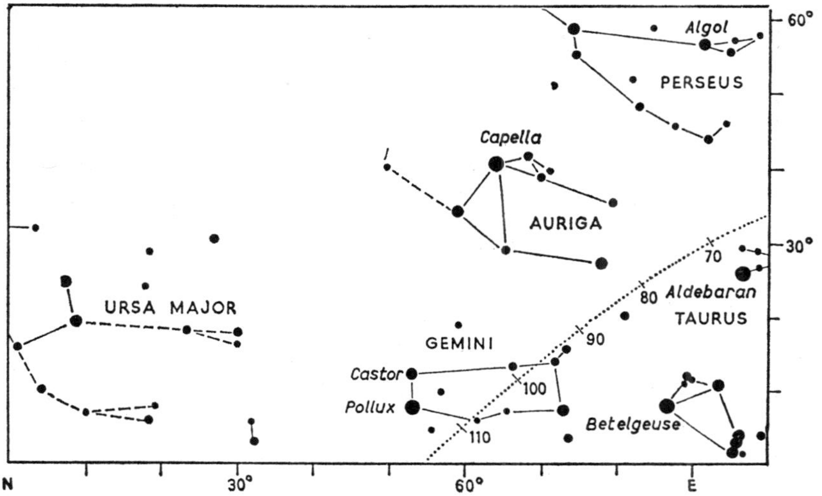

1977 YEARBOOK OF ASTRONOMY

11L

September 6 at 3ʰ	September 21 at 2ʰ
October 6 at 1ʰ	October 21 at midnight
November 6 at 23ʰ	November 21 at 22ʰ
December 6 at 21ʰ	December 21 at 20ʰ
January 6 at 19ʰ	January 21 at 18ʰ

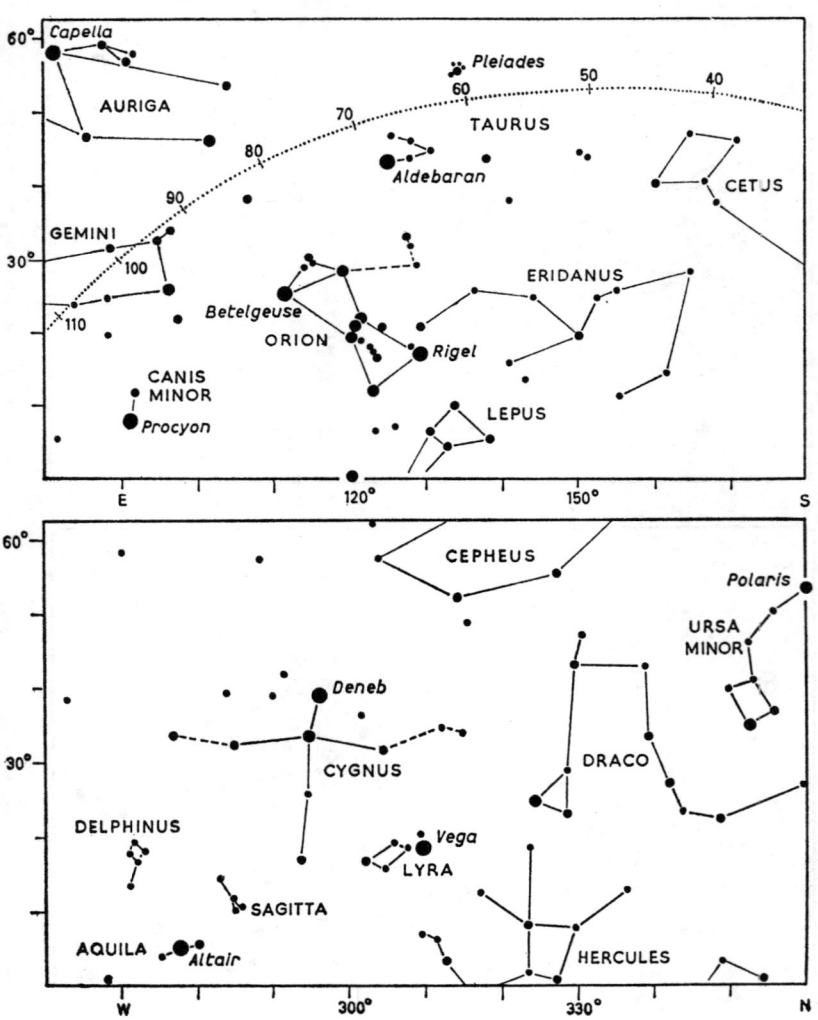

THE STAR CHARTS

September 6 at	3^h	September 21 at	2^h
October 6 at	1^h	October 21 at midnight	
November 6 at	23^h	November 21 at	22^h
December 6 at	21^h	December 21 at	20^h
January 6 at	19^h	January 21 at	18^h

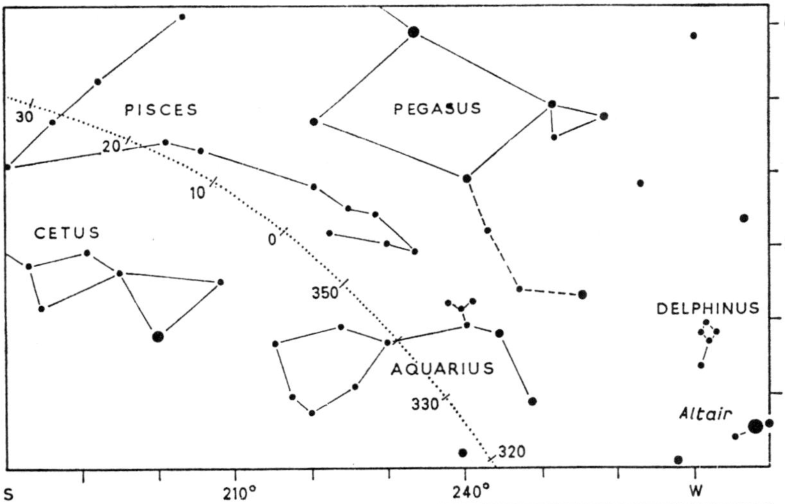

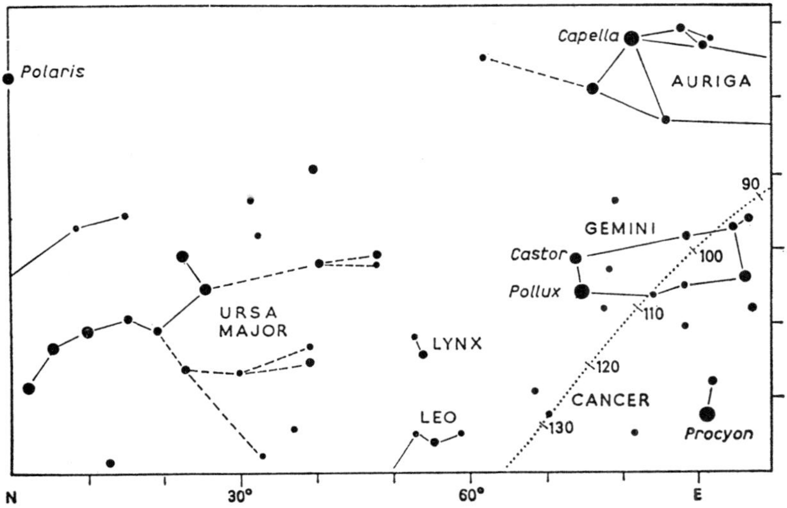

11R

12L

October 6 at 3^h	October 21 at 2^h
November 6 at 1^h	November 21 at midnight
December 6 at 23^h	December 21 at 22^h
January 6 at 21^h	January 21 at 20^h
February 6 at 19^h	February 21 at 18^h

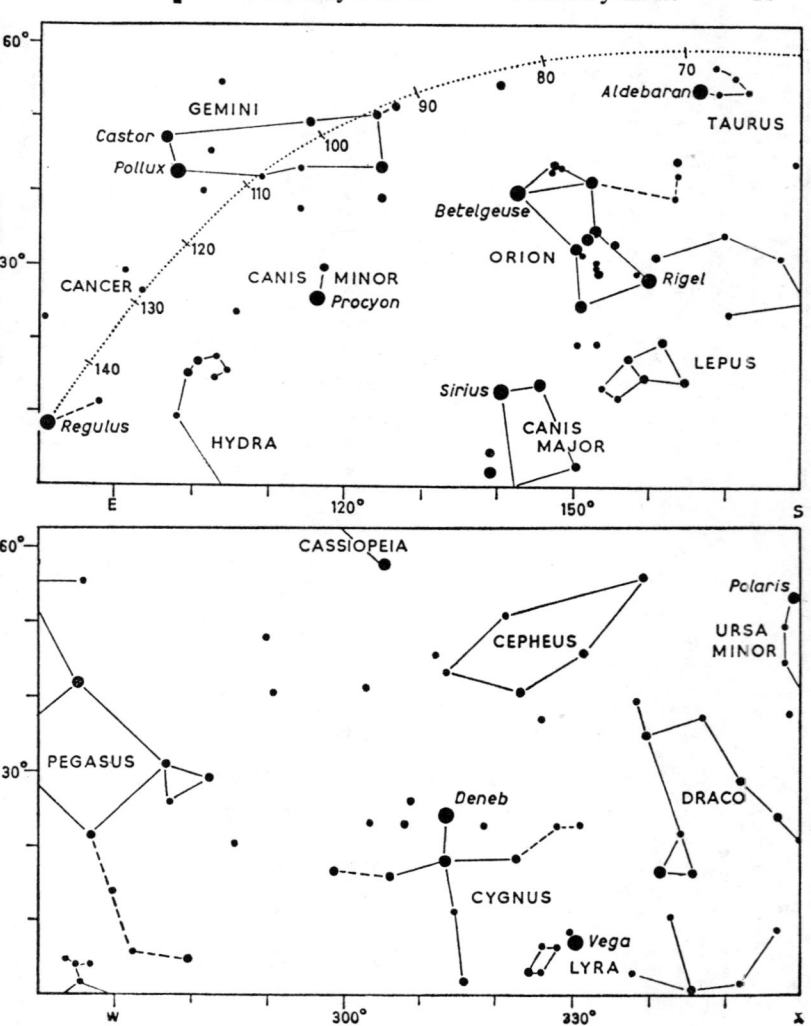

THE STAR CHARTS

October 6 at 3ʰ October 21 at 2ʰ
November 6 at 1ʰ November 21 at midnight
December 6 at 23ʰ December 21 at 22ʰ
January 6 at 21ʰ January 21 at 20ʰ
February 6 at 19ʰ February 21 at 18ʰ

12R

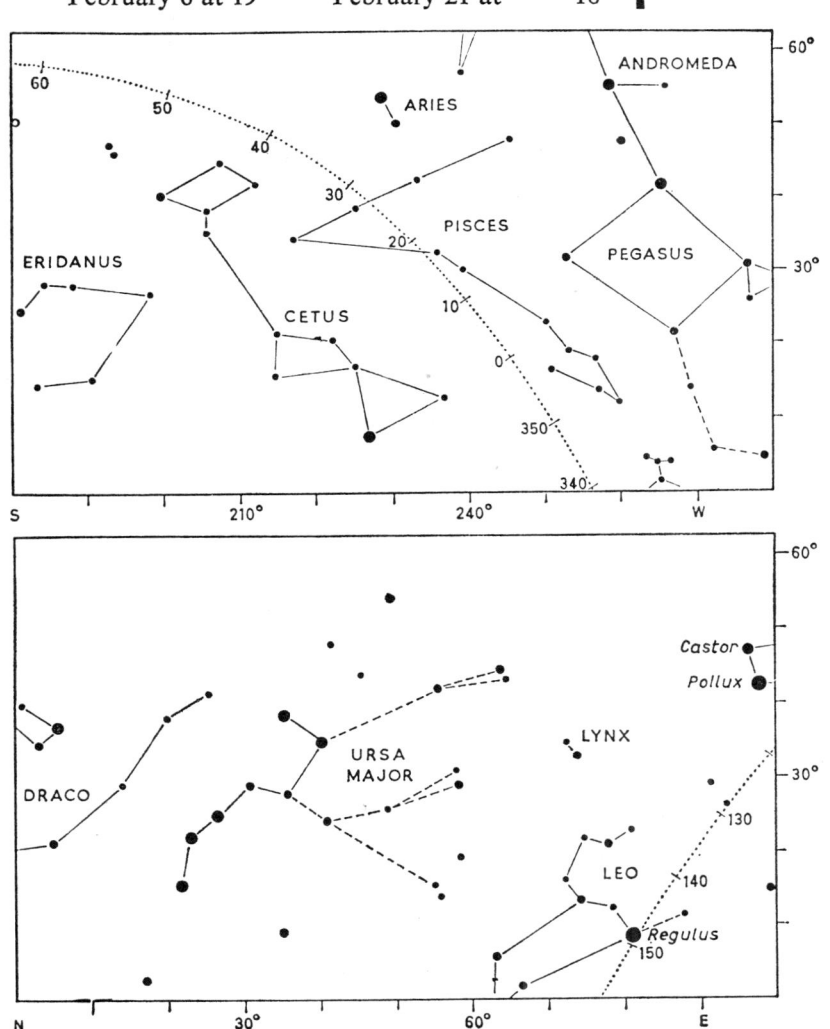

1977 YEARBOOK OF ASTRONOMY

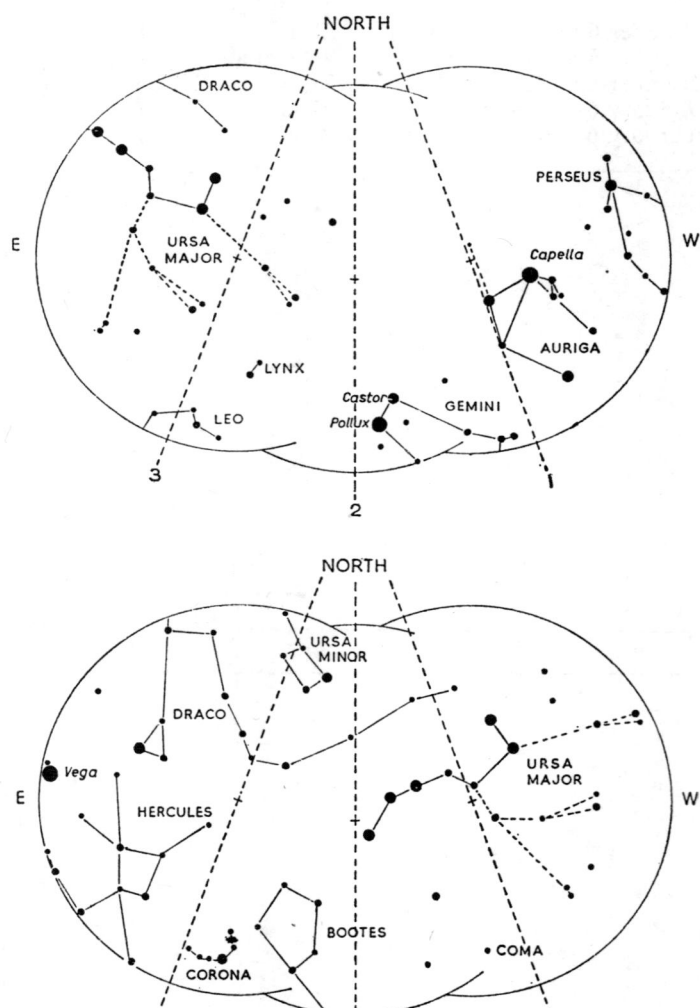

Overhead stars

THE STAR CHARTS

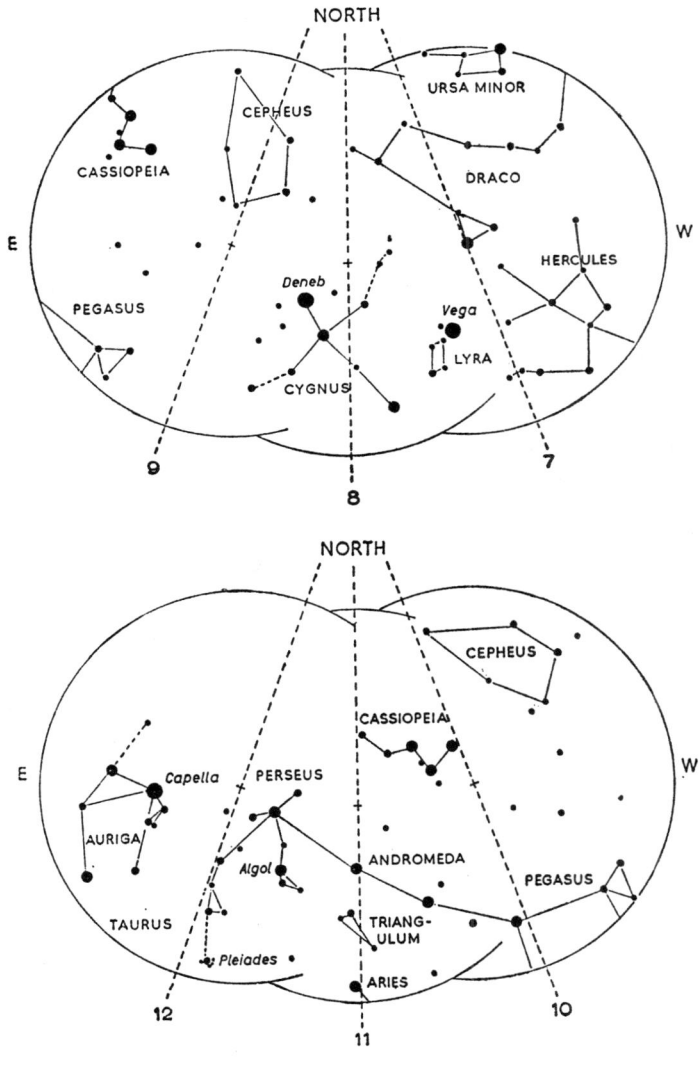

Overhead stars

Events of 1977

ECLIPSES

In 1977 there will be three eclipses, two of the Sun and one of the Moon. There will also be a penumbral eclipse of the Moon.

4 April – a partial eclipse of the Moon, visible in Europe, Africa, and America.

18 April – an annular eclipse of the Sun, visible in Africa, the Antarctic, and southern Asia.

12 October – a total eclipse of the Sun, visible in North America and the West Indies.

27 September – a penumbral eclipse of the Moon.

THE PLANETS

Mercury may best be seen in northern latitudes as an evening star near its greatest eastern elongation on 10 April, and as a morning star at greatest western elongation on 21 September.

Venus will be an evening star at the beginning of the year, reaching eastern elongation on 24 January and greatest brilliancy on 1 March. After inferior conjunction on 6 April it will be a morning star at greatest brilliance on 11 May and western elongation on 15 June.

Mars will be a morning star for the first half of the year, but rises at midnight by the end of July. It grows brighter at the end of the year as it approaches opposition in January 1978.

Jupiter is in conjunction on 4 June and reaches opposition on 23 December in Taurus.

Saturn is at opposition on 2 February on the borders of Cancer and Leo, and comes to conjunction on 13 August.

Uranus is at opposition on 30 April in Libra.

Neptune is at opposition on 5 June in Ophiuchus.

Pluto is at opposition on 2 April in Virgo.

The Planets in 1977

DATE		Venus	Mars	Jupiter	Saturn	Uranus	Neptune
January	6	332°	274°	52°	136°	221°	255°
	21	348	285	51	135	222	255
February	6	3	297	52	133	222	256
	21	14	309	54	132	222	256
March	6	20	319	55	131	222	256
	21	21	331	58	131	221	256
April	6	13	344	61	130	221	256
	21	7	355	64	130	220	256
May	6	9	7	67	131	220	256
	21	18	18	70	132	219	255
June	6	31	30	74	133	219	255
	21	45	41	78	134	218	255
July	6	60	52	81	136	218	254
	21	76	62	84	138	218	254
August	6	94	73	87	140	218	254
	21	111	83	90	142	218	254
September	6	130	93	93	144	219	254
	21	148	102	94	146	220	254
October	6	167	110	96	147	220	254
	21	186	118	96	149	221	254
November	6	205	124	96	150	222	255
	21	224	129	95	151	223	255
December	6	243	132	93	151	224	256
	21	261	132	91	151	225	256
Conjunction:		Apr. 6	—	June 4	Aug. 13	Nov. 4	Dec. 8
Opposition:		—	—	Dec. 23	Feb. 2	Apr. 30	June 5

43

Mercury moves so quickly among the stars that it is not possible to indicate its position on the star charts at a convenient interval. The monthly notes must be consulted for the best times at which the planet may be seen.

The positions of the other planets are given in the table on the previous page. This gives the apparent longitudes on dates which correspond to those of the star charts, and the position of the planet may at once be found near the ecliptic at the given longitude.

Examples:
(1) *What is the bright planet seen in the west near the Pleiades at* 21^h *in early April?*
Star chart 3L shows the stars in the western sky at this time, and since the planet is near the Pleiades it must have longitude about 60°. The table on page 43 identifies the planet as Jupiter.

(2) *Where may the planets Mars, Jupiter and Saturn be found at midnight on Christmas Eve, 1977?*
From the table on page 43, the longitudes of these planets for the nearest date December 21 are: Mars 132°, Jupiter 91° and Saturn 151°. Star chart 1L is used in this case and shows that Jupiter is high in the sky due south at the western end of the figure of Gemini; Mars is in the barren regions of Cancer, midway between Leo and Gemini; and Saturn is quite close to the star Regulus in Leo.

The Planets and the Ecliptic

The paths of the planets about the Sun all lie close to the plane of the ecliptic, which is marked for us in the sky by the apparent path of the Sun among the stars, and is shown on the star charts by a broken line. The Moon and planets will always be found close to this line, never departing from it by more than about 7 degrees. Thus the planets are most favourably placed for observation when the ecliptic is well displayed, and this means that it should be as high in the sky as possible. This avoids the difficulty of finding a clear horizon, and also overcomes the problem of atmospheric absorption, which greatly reduces the light of the stars. Thus a star at an altitude of 10 degrees suffers a loss of 60 per cent of its light, which corresponds to a whole magnitude; at an altitude of only 4 degrees, the loss may amount to two magnitudes.

The position of the ecliptic in the sky is therefore of great importance, and since it is tilted at about $23\frac{1}{2}$ degrees to the equator, it is only at certain times of the day or year that it is displayed to the best advantage. It will be realized that the Sun (and therefore the ecliptic) is at its highest in the sky at noon in midsummer, and at its lowest at noon in midwinter. Allowing for the daily motion of the sky, these times lead to the fact that the ecliptic is highest at midnight in winter, at sunset in the spring, at noon in summer and at sunrise in the autumn. Hence these are the best times to see the planets. Thus, if Venus is an evening star, in the western sky after sunset, it will be seen to best advantage if this occurs in the spring, when the ecliptic is high in the sky and slopes down steeply to the north-west. This means that the planet is not only higher in the sky, but will remain for a much longer period above the horizon. For similar reasons, a morning star will be seen at its best on autumn mornings before sunrise, when the ecliptic is high in the east. The outer planets, which can come to opposition and are then in the south at midnight, are best seen when opposition occurs in the winter months. Clearly the summer is the least favourable time to observe the planets, for the ecliptic is always low in the sky on summer nights.

Notes on the Planets in the monthly diagrams

The following general notes on observing the planets are followed by detailed month-by-month accounts of the behaviour of the planets, and of other interesting phenomena. These monthly notes include diagrams of the apparent movements of the planets at favourable times of the year. Additional notes on other astronomical phenomena will be found on the following pages.

The inferior planets, Mercury and Venus, move in smaller orbits than that of the Earth, and so are always seen near the Sun. They are most obvious at the times of greatest angular distance from the Sun (greatest elongation), which may reach 28 degrees for Mercury, or 47 degrees for Venus. They are then seen as evening stars in the western sky after sunset (at eastern elongations) or as morning stars in the eastern sky before sunrise (at western elongations). The succession of phenomena, conjunctions and elongations, always follows the same order, but the intervals between them are not equal. Thus, if either planet is moving round the far side of its orbit its motion will be to the east, in the same direction in which the Sun appears to be moving. It therefore takes much longer for the planet to overtake the Sun – that is, to come to superior conjunction – than it does when moving round to inferior conjunction, between Sun and Earth. The intervals given in the following table are average values; they remain fairly constant in the case of Venus, which travels in an almost circular orbit. In the case of Mercury, however, conditions vary widely because of the great eccentricity and inclination of the planet's orbit.

47

		Mercury	Venus
Inferior conj.	to Elongation West	22 days	72 days
Elongation West	to Superior conj.	36 days	220 days
Superior conj.	to Elongation East	36 days	220 days
Elongation East	to Inferior conj.	22 days	72 days

The greatest brilliancy of Venus always occurs about a month *before* greatest western elongation (as a morning star), or a month *after* greatest eastern elongation (as an evening star). No such rule can be given for Mercury, because its distance from Sun and Earth can vary over a wide range.

Mercury is not likely to be seen unless a clear horizon is available; it is seldom seen as much as 10 degrees above the horizon in the twilight sky. In general, it may be said that the most favourable times for seeing Mercury as an evening star will be in spring, some days before greatest eastern elongation; in autumn it may be seen as a morning star some days after greatest western elongation.

Venus is the brightest of the planets, and may be seen on occasions in broad daylight. Like Mercury, it is alternately a morning and an evening star, and will be highest in the sky when it is a morning star in autumn, or an evening star in spring. Venus is seen to best advantage when it comes to greatest eastern elongation in June; it is then well north of the Sun in the spring months and is a brilliant object in the sunset sky over a long period.

The superior planets, which travel in orbits larger than that of the Earth, differ from Mercury and Venus in that they can be seen opposite the Sun in the sky. The superior planets are morning stars after conjunction with the Sun, rising earlier each day until they come to opposition. They will then be in the south at midnight, and visible all night. After opposition, they are evening stars, setting earlier each evening until they set in the west with the Sun at the next conjunction. The interval between conjunctions or between oppositions is greatest for Mars (over two years). At the time of opposition, the planet is nearest the Earth, and therefore at its brightest. This change in brightness is most noticeable

NOTES ON THE PLANETS IN THE MONTHLY DIAGRAMS

with Mars, whose distance from the Earth can vary considerably; the other superior planets are at such great distances that there is very little change in brightness from one opposition to another. The effect of altitude is, however, of importance, for at a December opposition the planet will be among the stars of Taurus or Gemini, and can then be at an altitude of more than 60 degrees in southern England. At a summer opposition, when the planet is in Sagittarius, it may only rise to about 15 degrees above the southern horizon, and so make a less impressive appearance.

Mars, whose orbit is appreciably eccentric, comes nearest to the Earth at an opposition at the end of August; it may then be brighter even than Jupiter, but rather low in the sky in Aquarius. These favourable oppositions occur every fifteen or seventeen years (1924, 1941, 1956, 1971), but in Britain the planet is probably better seen at an opposition in the autumn or winter, when it is higher in the sky. Oppositions of Mars occur at an average interval of 780 days, and during this time the planet makes a complete circuit of the sky.

Jupiter is always a bright planet, and comes to opposition a month later each year, having moved, roughly speaking, from one Zodiacal constellation to the next.

Saturn moves much more slowly than Jupiter, and may remain in the same constellation for several years. The brightness of Saturn depends on the aspect of its rings, as well as on the distance from Earth and Sun. The rings are now closing again, and the planet is therefore less bright at each successive opposition.

Uranus, *Neptune* and *Pluto* are hardly likely to attract the attention of observers without adequate instruments, but some notes on their present positions in the sky will be found in the April and June notes.

Phases of the Moon, 1977

	New Moon				First Quarter			Full Moon				Last Quarter		
	d	h	m		d	h	m		d	h	m		d	h	m
								Jan.	5	12	10	Jan.	12	19	55
Jan.	19	14	11	Jan.	27	05	11	Feb.	4	03	56	Feb.	11	04	07
Feb.	18	03	37	Feb.	26	02	50	Mar.	5	17	13	Mar.	12	11	35
Mar.	19	18	33	Mar.	27	22	27	Apr.	4	04	09	Apr.	10	19	15
Apr.	18	10	35	Apr.	26	14	42	May	3	13	03	May	10	04	08
May	18	02	51	May	26	03	20	June	1	20	31	June	8	15	07
June	16	18	23	June	24	12	44	July	1	03	24	July	8	04	39
July	16	08	37	July	23	19	38	July	30	10	52	Aug.	6	20	40
Aug.	14	21	31	Aug.	22	01	04	Aug.	28	20	10	Sept.	5	14	33
Sept.	13	09	23	Sept.	20	06	18	Sept.	27	08	17	Oct.	5	09	21
Oct.	12	20	31	Oct.	19	12	46	Oct.	26	23	35	Nov.	4	03	58
Nov.	11	07	09	Nov.	17	21	52	Nov.	25	17	31	Dec.	3	21	16
Dec.	10	17	33	Dec.	17	10	37	Dec.	25	12	49				

All times are G.M.T.

Reproduced, with permission, from data supplied by the Science Research Council.

MONTHLY NOTES, 1977
NORTHERN HEMISPHERE

January

Full Moon: 5 January *New Moon:* 19 January

Earth is at perihelion (nearest to the Sun) on 3 January at a distance of 91·4 million miles (147·1 million km).

Mercury is in inferior conjunction on 6 January, and comes to greatest western elongation (25°) on 29 January. For a few days at the end of the month it may be possible to catch a glimpse of Mercury, very low in the south-east just before dawn, but the planet is not very bright. Mars is in the same part of the sky, but is even fainter and rises later.

Venus is an evening star and is a brilliant object setting in a dark sky some four hours after sunset. The planet will be at greatest eastern elongation (47°) on 24 January. Magnitude −3·8 to −4·1.

Mars is now a morning star, but rises in the south-east only shortly before the Sun. This state of affairs continues until the early summer months. There is no opposition of Mars in 1977, but the planet will become a brilliant object by the end of the year.

Jupiter was at opposition last November, and is still a brilliant object in the evening sky. It will be seen in the south in the early evening, and sets north of west in the early morning hours. The planet is at a stationary point on 15 January on the borders of Aries and Taurus (see diagram on page 99), and after this date it begins to move direct. Magnitude $-2·3$ to $-2·1$.

Saturn is approaching opposition, and although it rises about three hours after sunset at the beginning of the month, the days lengthen quite rapidly at this season, and by the end of January, Saturn rises at sunset and will be in the south at midnight. The planet is moving retrograde on the borders of Cancer and Leo (see diagram in February notes, page 57) and is growing brighter (magnitude $+0·3$ to $+0·1$); it should be easily recognized in this rather barren part of the sky. A small telescope will show the famous rings and also the largest satellite Titan, which revolves about the planet in nearly sixteen days. Titan will be at greatest eastern elongation on the nights of 2 and 18 January, and at western elongation on the 9th and 25th.

Vesta is at opposition on 9 January, in Gemini, near the star Delta (magnitude $+3·5$). The path of Vesta at this time is shown on the diagram in the February notes. Although the magnitude of the planet at this opposition is $+6·6$, it should be possible to find it with a pair of binoculars. See notes on page 57.

Mars in 1777

In the story of Mars mapping there can be no doubt that the most celebrated year before the Space Age was 1877. The opposition was a very favourable one with Mars almost at perihelion and so at its closest to the Earth as well as the Sun. Asaph Hall discovered the two satellites, Phobos and Deimos (see the notes for August) and, of course, G. V. Schiaparelli drew attention to the straight, artificial-looking streaks which he named *canali*.

Yet it is interesting to look back a century further – to 1777, when planetary astronomy was very much in its infancy. Even the Moon had not been really accurately mapped, and the best chart in existence was that by the German astronomer, Mayer; it was reasonably good, but had been drawn to a very small scale. Jupiter's Red Spot had been recorded; the rings of Saturn and the phases of Venus were well known, and several satellites of the giant planets had been discovered. Io, Europa, Ganymede, and Callisto were known in Jupiter's system, and Titan, Iapetus, Rhea, Dione, and Tethys in that of Saturn.

The known Solar System ended at Saturn. It was not until 1781 that William Herschel, using a home-made telescope, discovered Uranus. And in 1777 only Herschel seems to have been making systematic observations of the markings on Mars, again with a reflector of his own construction. The focal length of the telescope was 20 feet, and Herschel generally used a magnification of 300.

Herschel began his observations of Mars on 8 April 1777, and made a drawing of the surface features. He made further sketches on 17, 26, and 27 April – and so far as we know, that was all, so that our knowledge of the appearance of Mars in 1777 depends entirely upon these few drawings, and we have very little to guide us. Herschel was unquestionably one of the greatest observers of all time, but his main work was in connection with the stars, and his Solar System studies were incidental – even though he leapt to fame by his discovery of Uranus, made during a more or less routine survey of that part of the sky.

In 1777 the surface features of Mars were not well known, and they had not been named; the only familiar object was the white polar cap, which Herschel understandably believed to be made up of ice or snow (a view which was still held right up to the time of Mariner 4, in 1965). Herschel's four drawings of 1777 show some of the dark markings, but none which can really be said to be identifiable, and one has to admit that his sketches are of historical interest only. However, he did his best to make an estimate of the planet's axial rotation or 'day', and he was not very wide of the mark. In 1779, using his observations made in that year together with the four drawings of 1777, Herschel gave a period of 24^h 39^m 21^s 67. Years later the great German astronomer Johann Mädler re-examined Herschel's observations, and used them to derive a

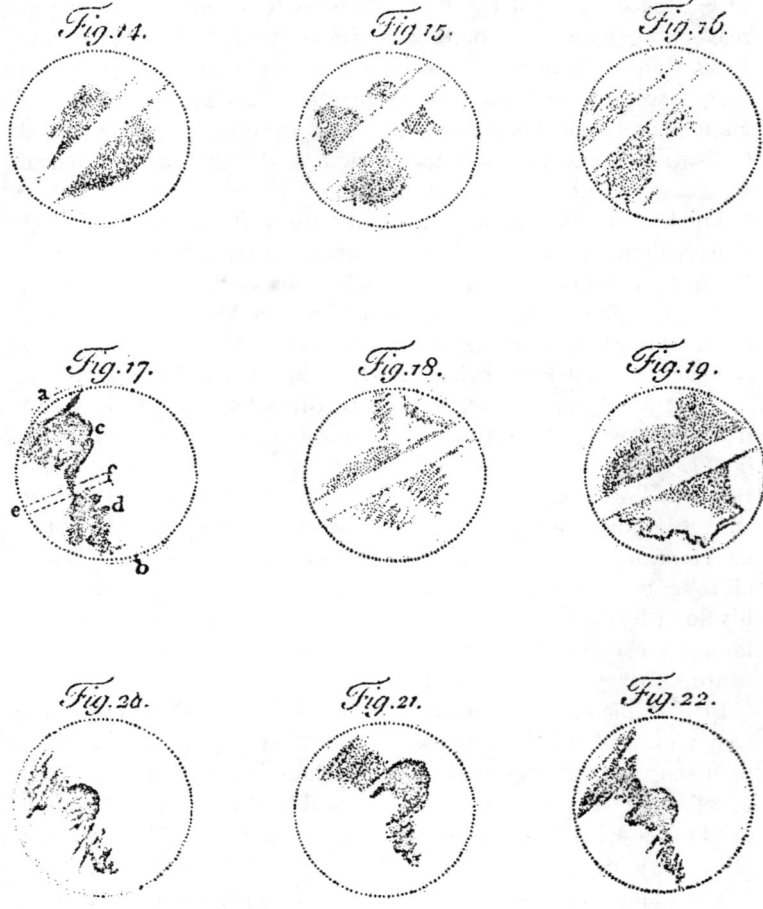

Drawings of Mars by Herschel, 1777

period of 24^h 37^m 26^s 27, which is less than four minutes in error according to modern determinations.

Herschel paid little further attention to Mars after 1781, and for the next few decades the main work was that of Johann Hieronymus Schröter, who concentrated upon planetary observation (and, incidentally, made extensive use of a telescope constructed by Herschel). However, it was not until the 1830s that Mädler and colleague Wilhelm Beer, produced the first reasonable map of Mars.

THE DIAMETERS OF THE ASTEROIDS

Vesta, at opposition this month, is the brightest of the asteroids, but it is not the largest; this honour must go to Ceres, the first-discovered member of the swarm (by Piazzi on 1 January 1801, the first day of the new century). Asteroid No. 2. Pallas, is also larger than Vesta; but the difference between the two is not so great as used to be thought.

All the asteroids appear so small in the sky that it is extremely difficult to measure their diameters by conventional methods. However, infra-red astronomy has come to the rescue (a method pioneered by D. A. Allen, one of our regular *Yearbook* contributors!), and it has been found that most of the asteroids are considerably larger than had been believed. Vesta was the first asteroid to be studied by Allen in this way.

It is of interest to compare the new, reliable values with those which were accepted until recently. The first ten asteroids are listed below; all diameters are given in miles.

Asteroid	Old value for diameter	New value (*infra-red*)
1 Ceres	427	650
2 Pallas	280	355
3 Juno	150	160
4 Vesta	240	335
5 Astræa	110	80
6 Hebe	105	140
7 Iris	90	130
8 Flora	80	105
9 Metis	135	130
10 Hygeia	220	260

Only in the case of Astræa is the infra-red value for the diameter smaller than the old value. Note that No. 3, Juno, is not third in order of size; it is much smaller than Hygeia. Other relatively large asteroids are 15 Eunomia and 16 Psyche (about 170 miles) and 511 Davida (190 miles). Another senior member of the swarm is 19 Fortuna (140 miles) which has a very low albedo or reflecting power.

Vesta owes its eminence to the fact that it is closer to the Sun than the other principal asteroids. Its mean distance from the Sun is 219,000,000 miles as against 257,000,000 miles for Ceres and Pallas and as much as 293,000,000 miles for Hygeia.

Capella at the Zenith

During evenings in January the brilliant yellow star Capella is almost at the zenith or overhead point as seen by observers in Great Britain or the northern United States. Capella is one of the brightest stars in the sky; it is markedly inferior only to Sirius, Canopus, and Alpha Centauri – and, of course, the two latter stars are in the far south, so that British observers and those in the New York area can never see them. Next in order come Arcturus, Vega, Capella, and Rigel, which are almost equal. These four are of contrasting colours; Capella's yellow hue is quite different from the pure white of Rigel, the steely blue of Vega and the lovely warm orange of Arcturus.

Capella is of spectral type G, similar to the Sun; but whereas the Sun is a Yellow Dwarf, Capella is a giant with about 200 times the Sun's luminosity. It is moreover, a binary, though the components are so close together that no telescope of amateur size will separate them or even show Capella as elongated in form.

February

Full Moon: 4 February   *New Moon:* 18 February

Mercury is a morning star, low in the south-east before dawn in the first few days of the month, but after this it will be too close to the Sun to be seen.

Venus grows a little brighter (magnitude $-4\cdot1$ to $-4\cdot3$) as it moves round to the near side of its orbit and comes closer to the Earth. Although this reduces its elongation from the Sun, it may still be seen in the evening sky for about four hours after sunset.

Mars is a morning star, but continues to rise only shortly before the Sun, and is not yet bright enough to be seen in the dawn sky.

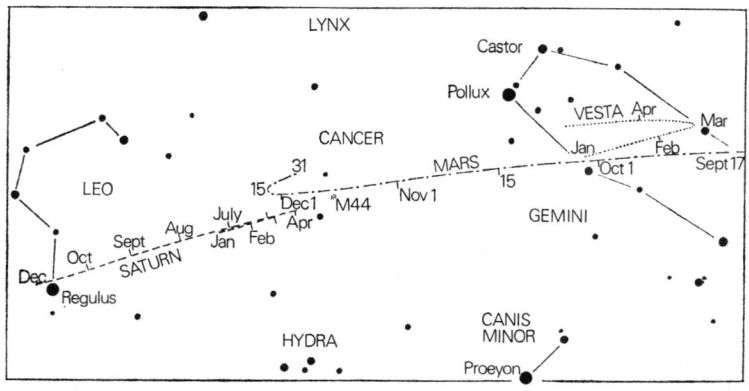

Saturn and Mars, 1977

Jupiter now moves into Taurus, forming a notable group with the

Pleiades and Hyades. The planet is somewhat less brilliant (magnitude $-2 \cdot 1$ to $-1 \cdot 9$) because of its increasing distance from the Earth, but it is easily found in the south shortly after sunset, and sets an hour or so after midnight.

Saturn is at opposition on 2 February in the constellation Cancer to the west of the Sickle of Leo. The rings continue to close, and the planet is not as bright (magnitude $+0 \cdot 1$) as it was at last year's opposition, although its distance at this time is about the same — 758 million miles (1220 million km). In the telescope the south pole of Saturn will be seen to be tilted towards the Earth. The satellite Titan is at eastern elongation on 2 and 18 February, and at western elongation on the 10 and 26.

J. J. Cassini

18 February is the tercentenary of the birth of a famous Italian astronomer, son of an even more famous father; Giacomo Cassini. usually known by the French version of his name – Jacques Cassini. His father, G. D. Cassini, became Professor of Astronomy at Bologna, and made notable studies of the planets; for instance, he discovered the gap in Saturn's rings which now bears his name, and he also found four of Saturn's satellites (Iapetus, Rhea, Dione, and Tethys). From 1668 he lived in France, and became a naturalized Frenchman five years later; he was given the official supervision of the new Paris Observatory, and came here at the express invitation of Louis XIV.

Jacques was born in Paris, and spent all his life in France. He succeeded his father at the Observatory, and was Director in all but name (there was no official Director until 1771). His work in connection with the size and rotation of the Earth was outstanding; he continued the planetary observations, and made one particulary interesting suggestion in connection with Saturn. The ring system had been known since Christiaan Huygens' observations in the 1650s, but its nature was problematical. Cassini believed that it might well be made up of small particles, each of which moved round the planet in an independent orbit. Cassini put forward this idea in 1715; not for over a century was it confirmed by the theoretical work of Clerk Maxwell.

Cassini also measured the proper motions of several bright stars, and attempted to measure the apparent diameter of Sirius, though – not surprisingly! – he had no success. He was an effective controller of the Observatory, and managed to obtain several new instruments for it. He died on 11 April 1756.

Of his three sons, one, César François or Cassini de Thury (1714-84) became an astronomer; in 1771 he became the first official Director of the Paris Observatory. The fourth Cassini, Jean Dominique (1748-1845) followed the family tradition at the Observatory, but his scientific career was cut short by the French Revolution; he was arrested as an aristocrat, and narrowly escaped execution. On his release he retired to the country, where he lived quietly until his death at the advanced age of ninety-seven.

Pallas

Another senior asteroid is at opposition this month: Pallas, on 10 February. Its magnitude is 6·5, so that binoculars will show it. However, it is very badly placed for northern-hemisphere observers, because its exceptionally high orbital inclination (35°) means that it can move well away from the Zodiac; and at present it lies in the southern constellation of Pyxis, below Hydra. Some smaller asteroids have orbital inclinations which are greater even than this. By contrast, the principal planets of the Solar System have low inclinations – 17° for Pluto and less than 10° for all the rest.

Comets, of course, may have steep inclinations, and many known comets have retrograde motion; but there is no known case of a retrograde asteroid.

The Closing Rings of Saturn

During the early 1970s Saturn has been a glorious sight, since its ring-system has been well displayed. Now, the rings are starting to close again, and by the end of the decade will return to the edge-on position last seen in 1966. Obviously, the effect is due only to the angle at which we see the rings.

It is worth noting that observers on the inner satellites of Saturn would never see the rings well, because all these satellites move more or less in the plane of the ring-system – so that the rings would always appear edgewise-on. The only satellites with appreciable orbital inclinations are the two outermost: Iapetus, which has an inclination of almost 15°, and the retrograde Phœbe. However, both these satellites are a long way from Saturn; Phœbe takes 550 days to complete one revolution.

THE COLOUR OF SIRIUS

Sirius, the brightest star in the sky, is at its best during February evenings so far as northern-hemisphere observers are concerned. It is a pure white star with an A-type spectrum, but because of its brilliance, and because it is always rather low down from the latitude of Britain, it seems to twinkle violently and to flash different colours. Oddly enough, Ptolemy and other observers of ancient times described it as being red. A change in colour has been suggested, but this seems to be most improbable.

Sirius is 8·6 light-years away, and is 26 times as luminous as the Sun. The tiny White Dwarf companion has 1/10,000 the luminosity of the primary; it is difficult to see because it is so overpowered, and small telescopes will not show it.

March

Summer Time in Great Britain and Northern Ireland commences on 20 March.

Full Moon: 5 March   *New Moon:* 19 March

Equinox: 20 March

Mercury is in superior conjunction on 16 March, and will not be visible until the very end of the month, when it begins to appear as an evening star (see April notes).

Venus reaches its greatest brilliancy (magnitude $-4\cdot3$) on 1 March and is then a splendid object setting in the late evening in a dark sky. During March the planet draws rapidly in towards the Sun and its brightness diminishes to magnitude $-3\cdot4$. By the end of the month it sets about an hour after the Sun.

Mars is still unfavourably placed, well south of the Sun and rising in the south-east less than an hour before sunrise.

Jupiter continues to be seen as an evening star, setting at midnight at the beginning of the month. The planet is moving direct in Taurus, and at the end of March will be seen south of the Pleiades (see diagram on page 99). Magnitude $-1\cdot9$ to $-1\cdot7$.

Saturn is still visible for most of the night, setting an hour or so before sunrise to the north of west. The planet is moving retrograde in Cancer, and as its distance increases its magnitude diminishes from $+0\cdot1$ to $+0\cdot3$. The satellite Titan is at eastern elongation on 6 and 22 March, and at western elongation on the 14th and 30th.

Venus at its Brightest

Venus is by far the brightest of all the planets, and at its best it can cast a perceptible shadow. This month it will be seen against a dark background, and will be truly magnificent. It is easy to understand why it was named after the mythological Goddess of Beauty.

Rather surprisingly, Venus is brightest when it is in the crescent phase. When more of the sunlit hemisphere is turned toward the Earth, Venus is further away; at full, of course, Venus is on the far side of the Sun (superior conjunction) and is to all intents and purposes out of view.

Venus is so brilliant partly because of its nearness – it is the closest of the planets to the Earth – and partly because of its high albedo or reflecting power. Venus reflects 76 per cent of the sunlight which reaches it, as against a mere 16 per cent for Mars. The albedoes of the principal planets are as follows:

    Mercury, 6 per cent
    Venus, 76
    Mars,   16
    Jupiter, 73
    Saturn, 76
    Uranus, 93
    Neptune, 84
    Pluto,   14

The value for Pluto is very uncertain. The Earth's albedo, as determined by measurements from space vehicles, is 36 per cent; that of the Moon is 7 per cent, showing that the Moon's rocks are extremely dark rather than being mirror-like.

Sir James Jeans, 1877-1946

One of Britain's great astronomers, Sir James Hopwood Jeans, was born a hundred years ago this month: on 24 March 1877, in

London. In addition to his pioneer research in many fields of astronomy, he was also one of the most skilful of all 'popularizers', and there are many people who remember his broadcasts.

Jeans was educated at Cambridge, and became Lecturer in Mathematics at the university in 1904. Later he spent four years in America, as lecturer at Princeton University, before coming back to England permanently. He first became well known in connection with his researches upon the past history of the Solar System, and it was he who developed the famous 'tidal theory' of the origin of the Earth and planets, which was generally accepted for many years even though it is now known to be incorrect.

The tidal theory involved the gravitational pull of a star which once by-passed the Sun. It was suggested that the passing star pulled a cigar-shaped 'tongue' of matter away from the Sun, and that after the star had moved away this 'tongue' broke up into drops, each drop becoming a planet. The largest planets, Jupiter and Saturn, were formed from the thickest part of the cigar-shaped 'tongue'.

Jeans did not originate the tidal theory; this had been done years earlier by Chamberlin and Moulton, in the United States. However, it was Jeans who perfected it. Later mathematical analyses showed that there were serious weaknesses in it, and today all tidal hypotheses have been abandoned. It is also worth noting that if Jeans's theory had been correct, planetary systems would have been extremely rare; the stars are so widely spread-out, even in the most crowded parts of the Galaxy, that close approaches can hardly ever occur. It is now believed that no second star is involved in the production of a system of planets, and such systems are almost certainly very common indeed.

Jeans also paid great attention to the problems of the internal constitution of the stars, and his name is often coupled with that of another famous pioneer, Sir Arthur Eddington, though in fact their viewpoints were decidedly different. Jeans wrote many technical papers, and several books; but later in his career he concentrated more and more upon popular writing. He was Secretary of the Royal Society for ten years (1919–29) and was knighted in 1928; subsequently he became Professor of Astronomy at the Royal Institution.

Of his popular books, two are particularly well remembered: *The Mysterious Universe* and *The Universe Around Us*. Jeans was a magnificent writer with a fluent, easy style and a knack of making difficult problems sound easy – though as a true scientist he was always very careful not to sacrifice accuracy by over-simplification. During the 1930s he made many broadcasts over the B.B.C. radio networks (in those days the National Programme and the various Regional wave-lengths) and his voice became very familiar all over Britain. It is not always easy now to recall that in that period sound radio was as influential as television is in the 1970s; and Jeans used it with the utmost skill.

He continued to broadcast upon astronomical matters during the war, and indeed until very near the time of his death in 1946. His Christmas lectures at the Royal Institution were among the most popular ever delivered there, and he also gave special lectures for children. He and Eddington (also a great popularizer) died within a few years of each other, and left a gap which has yet to be adequately filled.

Jeans was a man of varied interests; he was, for instance, a skilled musician – but probably he would like to be remembered by the description that he gave of himself. 'I am,' he said, 'a publicity man for the planets.'

Experiments with Time

Everyone is familiar with Summer Time, but how many people now remember Double Summer Time? For a period during the war, all clocks in Britain were kept one hour ahead of G.M.T. during the winter and two hours ahead during the summer, with a considerable saving in fuel – few people had to switch on their house lights before going to bed! Much more recently the experiment of keeping Summer Time throughout the year was tried but after two years a free vote in the House of Commons meant a reversion to the old system. Summer Time this year is in force between 20 March and 23 October.

April

Full Moon: 4 April   *New Moon:* 18 April

Mercury is at greatest eastern elongation (19°) on 10 April, and will then be favourably placed as an evening star. The diagram below shows the changes in altitude and azimuth (true bearing from the north through east, south, and west) of Mercury on successive evenings when the Sun is six degrees below the horizon; this is about 40 minutes after sunset at this time of year. The

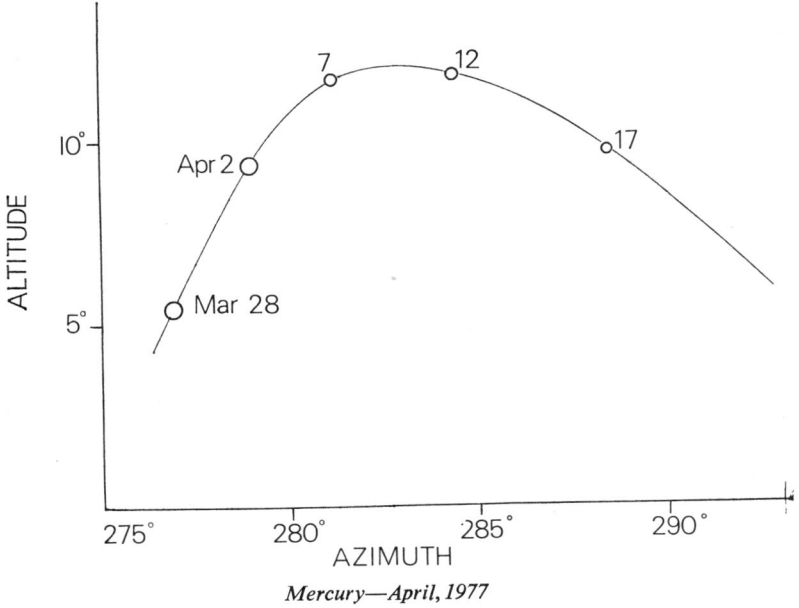

Mercury—April, 1977

changes in brightness are roughly indicated by the size of the circles, and it will be seen that Mercury is brightest *before* the date of eastern elongation.

Venus is in inferior conjunction with the Sun on 6 April. After this date the planet begins to appear as a morning star and will be seen at sunrise for the rest of the year. Venus moves out quite rapidly from the Sun, and by the end of April it rises almost due east more than an hour before sunrise.

Mars moves through Aquarius into Pisces during the month, but still rises less than an hour before sunrise. Magnitude $+1 \cdot 4$ to $+1 \cdot 3$.

Jupiter sets in the late evening to the north of west. During April the planet passes between the Pleiades and the Hyades, forming a striking group with these stars in Taurus. Magnitude $-1 \cdot 7$ to $-1 \cdot 5$.

Saturn is also an evening star and will be seen in the south at sunset. The planet reaches a stationary point on 11 April a few degrees east of the fourth-magnitude star Delta Cancri. Saturn is still a bright object, but is now fading as its distance from the Earth increases (magnitude $+0 \cdot 3$ to $+0 \cdot 5$).

Uranus is at opposition on 30 April on the western borders of Libra, some degrees from the third-magnitude star Alpha Libræ (Zuben-al-Genubi). The path of Uranus at this time is shown in the diagram in the May notes (page 71). At opposition, Uranus is 1630 million miles (2630 million km) from the Earth, and its magnitude is then $+5 \cdot 7$ so that it should just be visible to the naked eye. In a small telescope it appears as a greenish disk.

Pluto is at opposition on 2 April in Virgo. Its distance from the Earth at opposition is 2746 million miles (4420 million km), only about 25 million miles greater than the opposition distance of

Neptune (see June notes). The very eccentric orbit of Pluto causes it to come nearer at each successive opposition.

A partial eclipse of the Moon on 4 April will be visible in the British Isles and in North and South America. See notes on page 126.

An annular eclipse of the Sun on 18 April will be visible along a path which begins in the South Atlantic Ocean and crosses Zambia and Tanzania to end in the Indian Ocean. A partial eclipse will be seen in South Africa. For further details see page 126.

THE CRATERS OF MERCURY

There is a legend that the great astronomer Copernicus never saw the planet Mercury, because of mists rising from the river Vistula – which runs near his home town of Toruń, in Poland. The story is almost certainly untrue (for one thing Copernicus spent some time in Italy, where the skies are clear and transparent), but Mercury is never very prominent, and there must be many people who have yet to glimpse it. There is a good opportunity this month. Remember never to sweep for Mercury with binoculars or a telescope until the Sun has set, as there is always a chance that the Sun will enter the field of view, with tragic results for the observer's eyesight.

When Mercury is located, it looks surprisingly bright; it is in fact brighter than most stars, and can equal Sirius. The trouble is that it can never be seen against a dark background, since even at favourable elongations it sets soon after the Sun (or, at morning apparitions, rises not long before the Sun). Before the flight of Mariner 10 we knew very little about the surface features.

The first serious attempts at mapping Mercury were made almost a century ago by G. V. Schiaparelli, the Italian observer who will always be associated with his drawings of the 'canali' of Mars. Schiaparelli decided to study Mercury during daylight, with both the planet and the Sun high in the sky. This involved a telescope equipped with accurate setting circles, and conditions were always difficult. Schiaparelli produced a map, and named some of the dusky areas which he recorded, but it was clear that his chart could be no more than approximate.

He was followed by E. M. Antoniadi, a Greek astronomer who spent most of his life in France and used the powerful 33-inch refractor at the Observatory of Meudon, near Paris. Antoniadi was a tireless and a skilled observer; he too studied Mercury during daylight, and he too compiled a map. Eventually he wrote a short book, which was published in Paris in 1934: *La Planète Mercure*. The book was a noble attempt at summarizing what was known about Mercury, and it remained the standard work up to the flight of Mariner 10.*

Expert though Antoniadi was, his map of Mercury does not correlate well with what we now know about the surface features; he was much less successful than with Mars, which is hardly surprising. It was generally believed that the Mercurian surface must be rather like that of the Moon, but from Earth there was no hope of seeing any craters or mountain ranges. Moreover, Antoniadi believed that the atmosphere of the planet was dense enough to support dusty material in suspension, and he recorded local veilings and obscurations which he described as 'more frequent and obliterating than those of Mars' – a conclusion which we now know to be wrong, since the Mercurian atmosphere is negligible.

Mariner 10 revealed a landscape which is astonishingly lunar; there are craters, peaks, valleys, and dark plains, of which the most striking has been named the Caloris Basin. Without space probes Mercury would remain a world of mystery. Even large telescopes will show little upon its surface, and small instruments will reveal nothing apart from the characteristic phase; Mercury is, after all, a small planet, and never comes much within 50,000,000 miles of us.

---

*Both Antoniadi's classic books have recently been translated into English: *The Planet Mercury* in 1974 and *The Planet Mars* in 1975 (distributed by David & Charles Ltd, Newton Abbot, Devon). They are still of unique historical interest.

Green Stars

Uranus, not far from the star Alpha Libræ this month, is a decidedly green planet; telescopically the colour of its pale disk is unmistakable, though with binoculars it looks like a star. It is the only planet to show a greenish hue (Neptune, which is in many ways similar to Uranus, is bluish). Green stars, too, are very rare – excluding the smaller components of double stars in which the primary is red, so that contrast effects have to be taken into consideration.

One such star is Antares in the Scorpion, which is well placed for observation during the early hours of the morning during

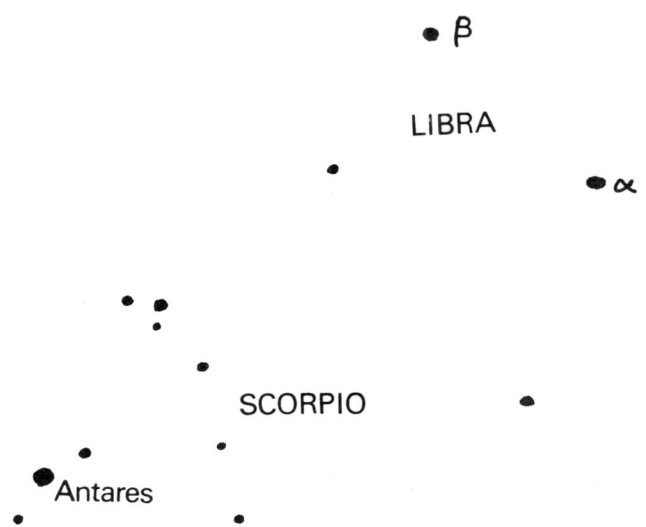

April and is an evening object later in the year. Antares itself is a red supergiant of spectral type M. Its companion shows a distinctly greenish tinge, and is interesting because it is one of those stars known to emit detectable radio waves. Rasalgethi or Alpha Herculis is another red supergiant with a companion which looks greenish; then too, there is Albireo or Beta Cygni, where the yellow primary is accompanied by a companion which is variously described as greenish or bluish. Beta Cygni is a wide, easy pair, and one of the most beautiful doubles in the sky. A small telescope will split it easily.

There is no known single star which is really green, but it is often said that there is a greenish cast to Beta Libræ, which has the cumbersome proper name of Zubenelchemale. Beta Libræ is of spectral type B, so that it is extremely hot. It is also very remote and luminous. It appears as a star of magnitude 2·7, but its absolute magnitude is −0·5, so that if it lay at the 'standard' distance of 10 parsecs (32·6 light-years) it would appear as the most brilliant star in the sky apart from Sirius and Canopus.

Beta Libræ is easy to find. No colour will be seen with the naked eye, and even with a telescope most people will describe it as white. However, it is worth looking at, and it is interesting to see whether any observers can detect a greenish cast in it.

May

Full Moon: 3 May   *New Moon:* 18 May

Mercury is at greatest western elongation (25°) on 27 May. The planet is then a morning star, but will be too low and not bright enough to be seen in the bright dawn sky.

Venus reaches its greatest brilliancy as a morning star on 11 May (magnitude $-4\cdot2$), and rises rather more than an hour before sunrise in the east. On 13 May Venus passes rather more than a degree north of Mars, but this planet is hardly bright enough to be seen at this time.

Mars is also a morning star, rising an hour or more before sunrise. It is moving direct through Pisces and is now beginning to grow brighter (magnitude $+1\cdot3$) and to move out from the Sun.

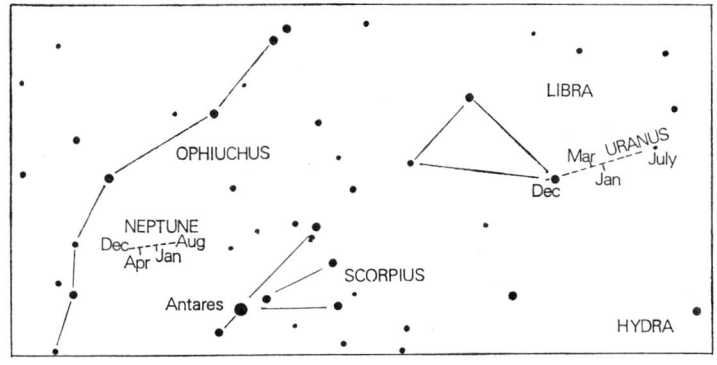

Uranus and Neptune, 1977

Jupiter is still visible as an evening star for about two hours after sunset at the beginning of May, but it is moving rapidly towards conjunction with the Sun. The planet is moving direct in Taurus and on 20 May it passes some degrees north of Aldebaran, but on that date it sets less than an hour after the Sun.

Saturn may be seen in the evenings in Cancer, moving direct towards Regulus and the Sickle of Leo. At the end of May it sets to the north of west at midnight. (Magnitude $+0\cdot5$ to $+0\cdot6$.)

Gamma Virginis – a Fine Binary Star

The constellation of Virgo is well placed for observation during May evenings. It is in the Zodiac, and it is one of the largest constellations in the sky. The leading star, Spica, is of the first magnitude, and the Y-form of Virgo is very easy to recognize.

There is one easy way to find Spica. Follow through the curve of the Great Bear's tail (or the handle of the Plough); this will lead first to the brilliant orange star Arcturus, in Boötes (the Herdsman) and then on to Spica.

The star at the base of the bowl of the Y, Gamma Virginis, is a particularly fine binary, and still a splendid sight in a small telescope, though admittedly it is less striking than it used to be a few decades ago. It has several proper names: Arich, Postvarta, and Porrima, though in general the names of stars below the first magnitude are seldom used.

Gamma Virginis is made up of two components, each of spectral type F and each of magnitude $3\cdot5$. They are, in fact, perfect twins, and it is impossible to tell which is the senior partner. The apparent separation at present is 5 seconds of arc, but this is becoming steadily less, and the year 2016 Gamma Virginis will appear single except in giant telescopes.

Obviously this does not mean that the components are really moving closer together; everything depends upon the angle from which we view them. The revolution period is approximately 180 years, and since the components are equal in mass the centre of gravity of the system is midway between the two. What is happening is that one component is moving 'behind the other' as seen from Earth, and by 2016 they will be more or less lined up, though the nearer component will not actually hide the more distant star.

There is another fine binary in Leo, which precedes Virgo in the Zodiac. This is Gamma Leonis, or Algieba, which is of the second magnitude, and therefore considerably brighter than Gamma Virginis. With Gamma Leonis the components are decidedly unequal, and the colours too are different; the primary is of type K, and is strongly orange, while the secondary has a G5-type spectrum. The present apparent separation is 4·3 seconds of arc, but this is slowly increasing; the revolution period is 407 years.

The existence of binary systems was first announced by Sir William Herschel. Previously it had been thought that all double stars must be optical pairs, due to chance alignments. Rather surprisingly, optical pairs are much less common than binaries, though of course many cases are known; one is Tau Virginis, which is made up of a fourth-magnitude star and a 9·5-magnitude companion at an apparent distance of 80 seconds of arc. With Tau Virginis the components are not genuinely associated, and merely lie in approximately the same direction as seen from Earth.

It used to be thought that a binary resulted from the break-up of an originally single star which rotated very rapidly, and became unstable; but this fission theory has now been generally abandoned, and it seems more likely that the components of a binary were produced in the same region of space from the same cloud of interstellar material.

R Coronæ

One of the most famous variable stars in the sky is well placed during evenings this month. This is R Coronæ, in the 'bowl' of the Northern Crown, not far from the brilliant Arcturus.

Corona is a small constellation, but its shape makes it easy to recognize, and it contains one fairly bright star, Alpha Coronæ, which is of the second magnitude. R Coronæ is generally just below naked-eye visibility, since its 'normal' magnitude is 6; also in the bowl of the Crown is a useful comparison star, which is of magnitude 6·6.

For most of the time R Coronæ remains approximately steady in light, with only slight fluctuations, but at any moment it may fade sharply, dropping down to below the 10th magnitude in only a few days. The recovery to maximum brightness is much slower, and accompanied by marked fluctuations. Minima can never be predicted; R Coronæ has been known to stay at maximum for more than ten consecutive years, but several times since 1973 it has fallen well below binocular visibility. It is not unique. Various similar stars are known (SU Tauri, in the Bull, is a good example), but R Coronæ variables are rare. Because of their unpredictability, they are favourite objects of study for amateur observers.

RECENT NOVÆ

The summer constellations of Aquila and Cygnus are now coming into view in the late evenings. Two interesting novæ have been seen in this region within the last few years. In 1967 the English amateur astronomer G. E. D. Alcock discovered a nova in Delphinus (the Dolphin), a conspicuous little group of stars not far from Altair. HR Delphini, as the nova is now called, reached magnitude 3½, and faded very slowly; it was indeed the slowest true nova on record, and at the beginning of 1976 was still above the 12th magnitude.

The second remarkable nova, V.1500 Cygni, blazed out in 1975, and rose rapidly to magnitude 1¾. It was the brightest nova to be well seen from Britain since DQ Herculis of 1934, but it faded very rapidly, and remained visible with the naked eye for only a few nights. For a time during its initial fading it was strongly red.

Novæ are commonest in the Milky Way zone, and several have been seen in Cygnus and in Aquila. Indeed, Nova Aquilæ 1918 was the brightest nova of the present century, and at its peak it surpassed all the stars apart from Sirius. It is still visible, but is now a faint telescopic object.

MONTHLY NOTES NORTHERN · JUNE

June

Full Moon: 1 June   *New Moon:* 16 June

Solstice: 21 June

Mercury is in superior conjunction on 30 June and will not be visible during the month.

Venus now rises about two hours before the Sun and is at greatest western elongation (46°) on 15 June. Although it is not quite as bright (magnitude −4·1 to −3·8) it is still a fine object in the morning sky. Venus passes about one degree south of Mars on 3 June, and it should be possible to see both planets on that morning.

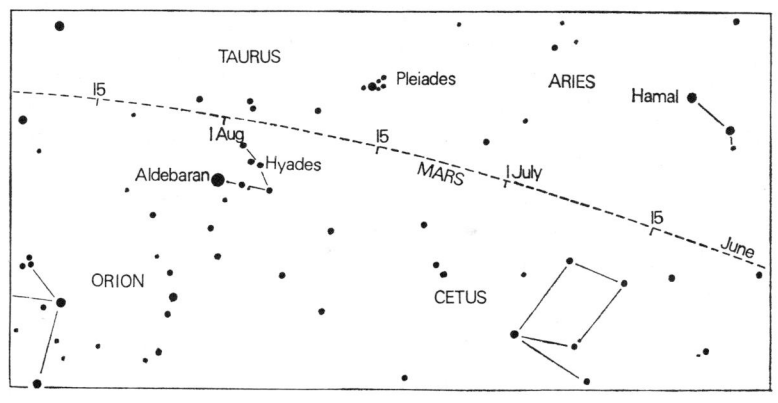

Mars, 1977

Mars now becomes more readily visible in the dawn sky, since it rises two hours or more before sunrise. It is growing a little brighter (magnitude $+1 \cdot 3$ to $+1 \cdot 2$) as its distance from the Earth decreases and is moving direct in Aries. Mars may be seen near Venus on the morning of 3 June and close to the crescent moon on 12 June.

Jupiter is in conjunction with the Sun on 4 June and then begins to appear as a morning star. By the end of the month it rises in the north-east an hour before the Sun.

Saturn is moving direct in Cancer, on the border of Leo, and sets in the late evening. Magnitude $+0 \cdot 6$ to $+0 \cdot 7$.

Neptune is at opposition on 5 June at a distance from the Earth of 2720 million miles (4380 million km). The planet is in the constellation Ophiuchus, which here intrudes into the Zodiac to the east of Antares. The position of Neptune is shown in the diagram on page 71. The planet is not visible to the naked eye (magnitude $+7 \cdot 7$) but may be found with binoculars or a small telescope.

OPHIUCHUS – INTRUDER INTO THE ZODIAC

The Zodiac is divided up into twelve constellations: Aries (the Ram), Taurus (the Bull), Gemini (the Twins), Cancer (the Crab), Leo (the Lion), Virgo (the Virgin), Libra (the Balance), Scorpio or Scorpius (the Scorpion), Sagittarius (the Archer), Capricornus (the Sea-goat), Aquarius (the Water-bearer) and Pisces (the Fishes) These constellations are very unequal in both size and importance; thus Scorpio and Gemini are large and brilliant, whereas Cancer, Libra and Pisces are decidedly obscure.

Originally the vernal equinox lay in Aries, which is why it is still called the First Point of Aries; precession has now shifted it into Pisces, and therefore Pisces should really be ranked as the first constellation of the Zodiac rather than the last. Similarly, the 'First Point of Libra', or autumnal equinox, is now in Virgo – roughly midway between the stars Beta and Eta Virginis.

It is often said that the Sun, Moon, and principal planets must remain within the boundaries of the twelve Zodiacal constellations, but this is not strictly true. Of course the planets do not keep strictly to the ecliptic, and one constellation which may contain planets is Cetus, which abuts on Pisces and Aries. There is, however, a constellation which is actually crossed by the ecliptic between Scorpio and Sagittarius. This is Ophiuchus, the Serpent-bearer (still occasionally referred to by its obsolete name of Serpentarius). Neptune is in Ophiuchus this month.

Ophiuchus is a very large constellation. It has no distinctive shape, and it adjoins another large, dim, group: Hercules – in fact Alpha Herculis and Alpha Ophiuchi lie fairly close together. Also in this region is Serpens, the Serpent with which Ophiuchus is apparently struggling. Broadly speaking, these three constellations lie in the area enclosed by imaginary lines joining Altair, Vega, Arcturus and Antares. Part of Ophiuchus lies in the northern hemisphere of the sky, but most of it is in the south, and it is the southernmost part which crosses the ecliptic between Scorpio and Sagittarius.

Alpha Ophiuchi or Rasalhague, is the brightest star in the constellation; it is of the second magnitude, with an A- type spectrum and a distance of 58 light-years. There are four more stars between magnitudes 2 and 3, but it must be admitted that there is a distinct paucity of interesting objects. There is, however, one fairly notable globular cluster, Messier 19, which lies about 8 degrees due east of Antares and is thus always rather low as seen from Britain and the northern United States. Binoculars will show it, and its form is easily seen with a small telescope. Its distance from us is 22,500 light-years, and unlike most globular clusters it is distinctly elliptical in form.

KARL FRIEDRICH GAUSS AND THE FIRST ASTEROID

In the history of mathematics one of the most honoured names is that of Karl Friedrich Gauss, inventor of the 'method of least squares'. He was born at Brunswick on 30 April 1777; he studied at Göttingen, and in 1807 became Director of the Observatory there. By then he had already established a great reputation – in a rather surprising way.

It had long been known that there is a wide gap in the Solar System between the orbits of Mars and Jupiter, and the relationship known as Bode's Law indicated that there should be an extra planet there. The story of Bode's Law has been told many times (really it should be the Titius-Bode Law, since it was first described by Titius) and even today nobody is certain whether or not it has any physical significance. However, in 1800, a group of astronomers met at the observatory in Lilienthal, near Bremen, to work out a plan for searching for the missing planet. Lilienthal Observatory was a private one, owned by Johann Schröter, the first great observer of the Moon; among those present were Dr Wilhelm Olbers, the famous German amateur astronomer, and the Baron Franz Xavier von Zach, of Hungary.

It was decided to 'parcel out' the Zodiac and carry out a systematic search for a new planet, which would certainly be faint. Letters outlining the proposal were sent to astronomers in various other countries, and one of these went to Giuseppe Piazzi, at the Observatory of Palermo.

Ironically, Piazzi had already been observing a suspicious looking object, which moved perceptibly from night to night and therefore could not be a star. He wrote to several astronomers about it, including Bode in Berlin, and the news came through to Schröter and von Zach; but Piazzi's letters took some time to arrive, and unfortunately he had not been able to make many observations of his moving object, partly because of illness and partly because of cloudy weather.

The object might have been lost – but in the summer of 1801 von Zach published an account of Piazzi's observations, and one man who read it was the young mathematician, Karl Gauss. Gauss thought that the observations were adequate for him to calculate an orbit – provided that he worked out an entirely new method of approach! This he did; he predicted where the object should be when it reappeared in the dawn sky, and he was correct. Using Gauss's prediction, Olbers rediscovered the object, which was, of course, the first asteroid: Ceres. (The name was given by Piazzi, since Ceres was the patron goddess of Sicily.) Continuing his observations, Olbers located a second asteroid in the same part of the sky; this was named Pallas. Two more were found within the next few years (Juno by Karl Harding, Schröter's assistant, and Vesta by the indefatigable Olbers). But the mathematical triumph was the work of Gauss. Until the end of his life in 1855 he produced many valuable papers, but astronomically he will always be particularly remembered for his connection with the rediscovery of Ceres.

GIOVANNI SANTINI

This is the centenary of the death of another astronomer who paid great attention to asteroids. Giovanni Santini was born at Caprese, Tuscany, in 1786; he became assistant at Padua Observatory in 1806, and succeeded to the Directorship in 1813. As well as his work in connection with comets and asteroids, he published a useful star catalogue in 1840. He died on 26 June 1877.

July

Full Moon: 1 and 30 July   *New Moon:* 16 July

Earth is at aphelion (farthest from the Sun) on 5 July, when its distance from the Sun will be 94·5 million miles (152·1 million km)

Mercury moves out from superior conjunction, and towards the end of the month it begins to appear as an evening star, but will be very low in the west in the twilight sky.

Venus now rises rather more than an hour after midnight. The planet is moving north, while the Sun moves south, and as a result Venus makes a more prolonged appearance in the morning sky. At the end of July Venus may be seen in the east for about three hours before the Sun rises. On 30 July, Venus passes less than two degrees south of Jupiter (magnitudes: Venus $-3·6$, Jupiter $-1·6$).

Mars is a morning star rising to the north of east shortly after midnight. The planet passes into Taurus, and in mid-July will be seen south of the Pleiades. By the end of the month Mars will form a notable group with the two clusters, the Pleiades and Hyades. The planet is growing brighter (magnitude $+1·2$) but will be seen to be not yet as bright as Aldebaran ($+0·9$). The rapid easterly motion of Mars, amounting to nearly three-quarters of a degree a day, will be readily noticed.

Jupiter now rises about two hours after midnight and will be seen in Taurus below the figure of Auriga. The planet is moving direct but is overtaken by Venus at the end of the month, as mentioned in the notes above.

Saturn now sets shortly after the Sun, and by the end of the month will be lost in the twilight.

The Moon – after Apollo

Eight years ago this month, the first two astronauts stepped out on to the surface of the Moon. Just before 3 a.m. on 21 July 1969, listeners all over the world heard Neil Armstrong's words: 'That's one small step for a man – one giant leap for mankind.' The gap between Earth and Moon had been bridged at last.

The idea of reaching the Moon was very old, and indeed it goes back to Classical times, but until recently there was no real prospect of achieving it. Various ideas were put forward, the most celebrated being the space-gun described in Jules Verne's famous novel, the first part of which was published in 1865; but even at the end of the last war, lunar travel was still regarded as a dream of the indefinite future. The situation changed dramatically in 1957, with the launching of Sputnik 1, the first artificial satellite. Two years later the first rockets reached the vicinity of the Moon; in October 1959 Luna 3 sent back the first pictures of the far side, which is never visible from Earth because it is always turned away from us. During the 1960s the whole of the Moon's surface was mapped by the U.S. Orbiter vehicles; soft landings with automatic vehicles were made proving that the lunar ground is firm enough to bear the weight of a space-craft; and the climax came with Apollo 11. That epic journey ended at 4.49 a.m. on 24 July 1969, when Armstrong, Aldrin, and Collins splashed down safely in the Pacific. Their time of arrival was a mere thirty seconds behind schedule.

Since then there have been six more Apollo flights, of which only one (Apollo 13) was unsuccessful. All the other expeditions were carried through with only minor hitches, and in each case specimens of lunar material were brought back for analysis. Studies of the 'Moon rocks' took a long time, and are not complete even yet, but a tremendous amount has been learned from them. The rocks are essentially basaltic, and they have shown that the Moon and the Earth are of approximately the same age (around

4,700 million years). There is no evidence of hydrated materials, proving that the lunar 'seas' were never water-filled, and – as expected – there was no sign of life, either past or present. The Moon has been sterile throughout its long history. Yet it is not completely inert, and minor tremors are common. These tremors are recorded by the seismometers left on the surface by the astronauts, and it is important to remember that apart from the instruments left by Armstrong and Aldrin, all the ALSEPS (Apollo Lunar Surface Experimental Packages) are still sending back data.

One notable result of the Apollo programme has been that the Moon is no longer the business of astronomers alone. Scientists of many other disciplines are concerned – particularly the geologists. It is also true to say that even now there are many problems to be solved, and the old argument about the origin of the craters (volcanic or meteoritic?) continues, even though it is now generally agreed that both processes must have operated.

We cannot yet tell when men will go back to the Moon. There may be a delay of more than ten years yet – perhaps even longer; but there seems little doubt that a Lunar Base will be set up in the foreseeable future. Such a Base should certainly be in existence before the end of the century. Without Apollo, anything of the sort would have remained a dream, and July 1969 will always be remembered as one of the most significant months in the history of mankind.

THE 'WILD DUCK'
One of the most beautiful galactic clusters in the sky is well placed for observation during July evenings. This is Messier 11 (NGC 6705), often known by its nickname of the 'Wild Duck'.

M. 11 lies in the small and otherwise unremarkable constellation of Scutum (the Shield), formerly known as Scutum Sobieskii or Clypeus Sobieskii – the shield of John Sobieski, the third King of Poland. The constellation was formed by Hevelius in the late seventeenth century; it lies below Aquila, and the best guides to it are the two stars Lambda and 12 Aquilæ. It is a rich area, immersed in the Milky Way, though Scutum itself contains no bright stars.

The discovery of M.11 is due to Kirch, in 1681, though the cluster is on the limit of naked-eye visibility; binoculars show it easily, and even a small telescope will bring out the characteristic fan-shaped form. Its distance is estimated at 5,500 light-years, and it contains about 500 stars brighter than the fourteenth magnitude. The star density is relatively high. If our Sun lay in the middle of the cluster, we would see between forty and fifty stars bright enough to cast shadows, and there would be no true darkness at night.

ANOTHER PLANETARY CONJUNCTION

The conjunction of Venus and Jupiter at the end of the month will be reasonably spectacular, and it will be interesting to follow the relative movements of the two planets for the mornings before and after the conjunction date on 30 July. Venus is much the more brilliant of the two, and is a full two magnitudes brighter than Jupiter.

Generally speaking, Venus and Jupiter are the two most imposing planets as seen with the naked eye. Venus is difficult to compare with anything else, because its superiority is so marked, and it is not often to be seen against a really dark background. Jupiter is always brighter than any of the stars, and at its best can attain magnitude $-2 \cdot 5$. Occasionally it is surpassed by Mars, whose maximum magnitude is $-2 \cdot 8$; but Mars remains at its best for only brief periods, and exceeds Jupiter only when near a perihelic opposition. At its faintest Mars sinks to about the second magnitude, and is easily confused with a star. At present it is comparable with Aldebaran, but by the end of the year it will be brighter than any star apart from Sirius and Canopus.

August

New Moon: 14 August   *Full Moon:* 28 August

Mercury is at greatest eastern elongation (27°) on 8 August, and is then an evening star, but is very low in the west at sunset and not very bright.

Venus rises to the north of east about three hours before sunrise. On 23 August it passes about six degrees south of Pollux, and as it rises in a dark sky, it should be possible to see its position in relation to the stars of Gemini. Venus now passes round to the far side of its orbit, and gradually loses its brilliance (magnitude $-3 \cdot 6$ to $-3 \cdot 5$) as its distance increases.

Mars is growing brighter (magnitude $+1 \cdot 1$) and now rises before

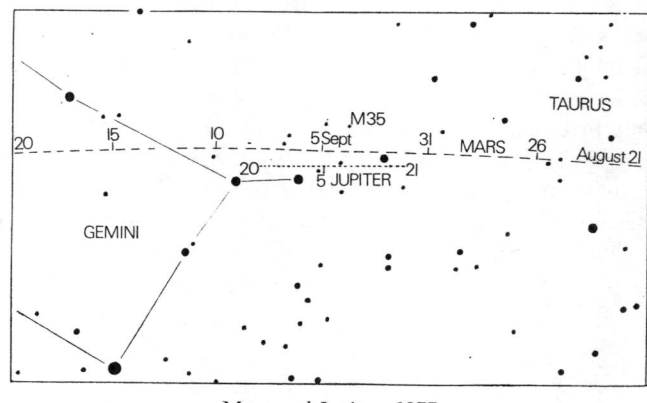

Mars and Jupiter, 1977

midnight to the north of east. On 1 August it passes five degrees north of Aldebaran (magnitude $+0 \cdot 9$) and continues to move through Taurus towards Gemini. The rapid eastward motion of Mars is now very obvious and by the end of August Mars is less than three degrees west of Jupiter. The two paths in the sky are very similar and could not be shown separately on the scale of the diagrams in the June and November notes. They are therefore shown below on a larger scale for the period 21 August to 20 September, and it will be seen that Mars passes close to Jupiter in early September.

Jupiter is very conspicuous (magnitude $-1 \cdot 6$ to $-1 \cdot 7$) to the north of Orion. It now rises at midnight and passes from Taurus into Gemini, but is rapidly overtaken by Mars, as shown in the diagram opposite.

Saturn is in conjunction with the Sun on 13 August and after this date it begins to appear as a morning star. By the end of August it rises about an hour before sunrise, but is not yet very bright (magnitude $+0 \cdot 7$).

ASAPH HALL AND THE SATELLITES OF MARS

Tennyson referred to 'the snowy poles of moonless Mars'. This was understandable enough; when he wrote that particular poem no attendants of Mars were known, and it was universally believed that the polar caps were snowy or at least icy in nature.

Swift, in Gulliver's *Voyage to Laputa*, had described how the astronomers of his remarkable flying island had discovered two Martian satellites, one of which moved round the planet in a period less than that of a Martian day; Voltaire, in *Micromégas*, had also written about two satellites – but at that time no telescope in existence was strong enough to show the dwarf attendants which were found in 1877 by Asaph Hall with the great new refractor at Washington. This had (and has!) an aperture of 26 inches, and was ideally suited to the task of searching for small satellites. Yet at first Hall had no success and he was about to give up the hunt when his wife persuaded him to have one more attempt.

On 11 August Hall detected a tiny point of light which, he thought, might well be a satellite; but mists rising from the Potomac River cut short his observations, and the next few nights were cloudy. Then, on 16 and 17 August, he was able to verify the original satellite and also to discover the second. We know them today as Phobos and Deimos – and thanks to Mariner 9, we even have detailed maps of their rough, cratered surfaces. They remain the smallest known satellites in the Solar System, and are irregular in shape, so that they are utterly unlike our own large, massive Moon.

There is no real mystery about the Swift and Voltaire descriptions, which were not, in any case, meant to be taken seriously. Venus had no satellite; the Earth had one; Jupiter, much farther from the Sun, had four – and so how could Mars possibly manage with less than two? However, it is pleasant to note that modern astronomers have given the two largest craters on Deimos the names of Swift and Voltaire!

During August 1977 Mars, in Taurus, may be compared with Aldebaran, the red star which is nicknamed 'the Eye of the Bull'. The two are not only of the same brilliancy, but also of very much the same colour. However, Mars is becoming steadily brighter, and will continue to do so until it reaches opposition in January 1978. This will not be a really favourable opposition, because Mars is some way from perihelion. At its best it will rival Sirius, the brightest star in the sky; but at a perihelic opposition (as in 1971) Mars may outshine Jupiter, so that it becomes the brightest natural object in the sky apart from the Sun, the Moon, and Venus. The two planets are close together in the sky at the end of the present month, but at the moment Jupiter is the brighter by well over a magnitude.

THE PERSEID METEORS

Many meteor showers occur annually, but there is no doubt that the August shower is the most consistent. The Perseids are very much in evidence from the last week in July right through to the third week in August, and an observer who stares upward at a dark, clear sky for a few minutes during that period will be very unlucky not to see at least one meteor. This year moonlight will not interfere after the first few days of August, so that the Perseids should be at their best.

THE 'SUMMER TRIANGLE'

All through the summer, the evening skies as seen from Europe and most of the United States are dominated by the three brilliant stars which make up what has been unofficially called the 'Summer Triangle'. These three stars are Vega or Alpha Lyræ, Altair or Alpha Aquilæ, and Deneb or Alpha Cygni.

The one thing that they have in common is spectral type; all are of class A – A0 for Vega, A7 for Altair, and A2 for Deneb, so that all are hot stars. Altair and Deneb appear white, but Vega has a distinct steely-blue cast which marks it out at once. Vega is also the most brilliant of the three; its magnitude is 0·04, and it is surpassed only by Sirius, Canopus, Alpha Centauri, and Arcturus. Altair is of magnitude 0·77, and Deneb appreciably fainter at 1·26.

Yet appearances are deceptive; and of the three Deneb is much the most luminous. According to one estimate it is about 10,000 times more powerful than the Sun, as against 50 Sun-power for Vega and only 9 for Altair. It is interesting to see what would happen if we could view them from the standard distance of 10 parsecs or 32·6 light-years. Deneb would then be of magnitude −7, so that it would be splendid indeed, and would cast strong shadows; but Vega would be slightly fainter than it actually appears (magnitude 0·5), and Altair would be reduced to magnitude 2·2.

Obviously, then Vega and Altair are relatively close by stellar standards. Altair is 16 light-years from us, so that in 1977 we see it as it used to be in 1961; Vega, at 26 light-years, appears in its 1951 guise. But Deneb is thought to be approximately 1,600 light-years away, so that we are seeing it as it used to be in the time of the Roman occupation of Britain!

This whole region of the sky is very rich, and is crossed by the

Milky Way, which is particularly striking in Cygnus. Lyra, which includes Vega, is a small constellation with more than its fair share of interesting objects; there are, for instance, the quadruple star Epsilon Lyræ, the eclipsing binary Beta Lyræ, and the Ring Nebula, M.57, one of the most striking of planetaries (easily found, as it lies midway between Beta and Gamma Lyræ). Cygnus is in the form of a cross, though the symmetry is spoiled by the fact that one member of the group is fainter and farther away from the centre than the rest; however, this star, Beta Cygni or Albireo, is the lovely coloured double. Aquila really does give a vague impression of a bird in flight; Altair is flanked to either side by a fainter star – a characteristic which it shares with Antares in Scorpio. However, there is no fear of confusing the two, because Antares is fiery red and is much lower down as seen by northern-hemisphere observers.

Eta Aquilæ, the centre member of the line of three naked-eye stars below Altair, is a Cepheid variable, with a period of 7·1 days. It appears as bright as the prototype star, Delta Cephei; and had it been studied first, no doubt these vitally important variables would now be known as Aquilids.

September

New Moon: 13 September   *Full Moon:* 27 September

Equinox 23 September

Mercury is in inferior conjunction on 5 September, but moves quickly out from the Sun to reach greatest western elongation (18°) on 21 September. In our northern latitudes this is the most favourable opportunity of seeing this elusive planet as a morning star. The diagram shows the changes in altitude and azimuth of Mercury on successive mornings when the Sun is six degrees

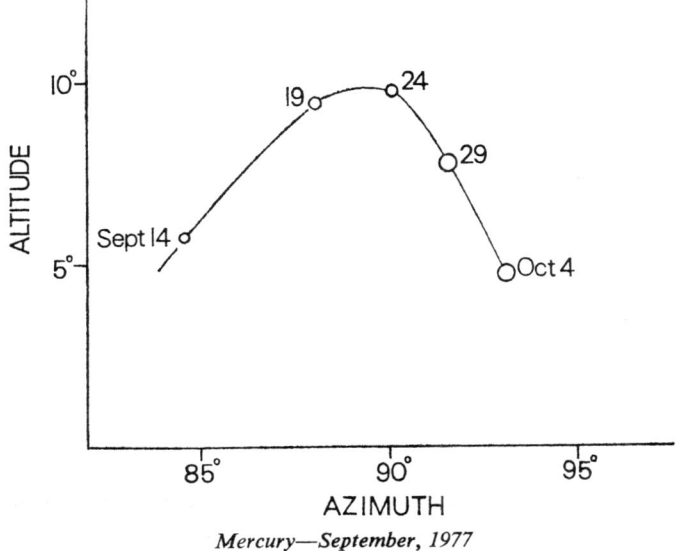

Mercury—September, 1977

below the horizon; this is about 35 minutes before sunrise in September. The changes in brightness are roughly indicated by the size of the circles, and it will be seen that Mercury is brightest *after* the date of greatest elongation.

Venus continues to dominate the morning sky, and rises nearly three hours before sunrise (magnitude -3.4). On 18 September it passes less than half a degree south of Saturn, and the change in positions on the 18th and 19th should be looked for as the planet rises in a dark sky. On 22 September, Venus passes less than half a degree north of Regulus.

Mars is now growing noticeably brighter (magnitude $+1.0$ to $+0.8$) and rises about an hour before midnight. The planet moves into Gemini and is not very far from the brighter Jupiter, so that at the beginning of the month the two planets rise at about the same time. Mars passes half a degree north of Jupiter on the night of 4 September, the closest approach occurring as the two planets rise to the north of east. Mars will be near the star Delta Geminorum at the end of September.

Jupiter moves direct in the western part of Gemini and rises in the late evening. The conjunction with Mars on 4 September is mentioned in the notes above, and it should be quite interesting to watch the rapid motion of Mars, which is about four times that of Jupiter at this time. Jupiter continues to grow brighter (magnitude -1.8 to -1.9) as the distance from the Earth decreases.

Saturn now rises before dawn and may be seen for a while in a dark sky. The planet is moving direct in Leo, some degrees to the west of Regulus (see diagram on page 57).

A penumbral eclipse of the Moon on 27 September will be visible in North America. See notes on page 126.

MONTHLY NOTES NORTHERN · SEPTEMBER

LE VERRIER – EXPLORER OF THE SOLAR SYSTEM

This month is the centenary of the death of one of the most famous of all French astronomers: Urbain Jean Joseph Le Verrier. Though he will always be best remembered for his discovery of the planet Neptune, he carried out much other astronomical work of outstanding importance.

Le Verrier was born at St. Lo, in Normandy, on 11 March 1811, and was educated in Paris. In 1839 he became a lecturer in astronomy at the École Polytechnique, and produced various mathematical papers which were well received. Then, in 1845, he began to investigate the irregularities in the movements of the planet Uranus; he thought – correctly – that these irregularities must be due to the gravitational influence of an unknown planet, and was able to predict where the new planet should be. He sent his results to the Berlin Observatory, and the planet was duly located by Johann Galle and his colleague, Heinrich d'Arrest. In England, Adams had made similar calculations, but through a series of delays and misunderstandings the planet was not identified until after the success at Berlin. The story of the discovery of Neptune has been told many times; both Adams and Le Verrier deserved the greatest possible credit, and are now recognized as co-discoverers, though it cannot be denied that the first positive identification was due entirely to the work of Le Verrier.

In 1853 Le Verrier became Director of the Paris Observatory, and two years later he began paying serious attention to the innermost part of the Solar System. The planet Mercury was not behaving quite as it might be expected to do; could there be yet another planet, this time extremely close to the Sun? Le Verrier believed so; but he also knew that the planet would be hard to see, and might be observable only when it passed in transit across the face of the Sun (as both Mercury and Venus sometimes do).

By 1860 Le Verrier was sufficiently confident to deliver a lecture in which he stated that Venus was much more massive than was generally thought, or that an intra-Mercurian planet really existed. Professor Rudolf Wolf, of Zürich, drew attention to some suspicious round spots which had been seen from time to time on the Sun, and which could possibly be due to the new planet; another astronomer, Dr Haase, provided some extra cases; and there was also a remarkable observation made by a French doctor named Lescarbault, on 26 March 1859. Lescarbault, using a small telescope and primitive timekeeping equipment, claimed to have followed a moving dot across the Sun's face for well over an hour.

Le Verrier was interested enough to go to Lescarbault's home at Orgères, where he interviewed the doctor – who was also the local carpenter. (He recorded his observations on planks of wood, planing them off when he had no further use for them.) Rather surprisingly, Le Verrier was convinced that the observation was valid, and he was led on to compute an orbit for the planet, which was then dignified with a name – Vulcan, after the mythological blacksmith of the gods. According to Le Verrier, Vulcan had a revolution period of 19 days 17 hours; it moved round the Sun at a mean distance of 13,000,000 miles, and it had a mass of 1/17 of that of Mercury, so that by planetary standards it was decidedly small.

The one hope of finding Vulcan except at a transit was at a total eclipse of the Sun, when for a brief period the sky becomes dark, and bright stars and planets shine out. There was a total eclipse in 1860, and Le Verrier organized a systematic hunt for Vulcan. Nobody found it. Moreover, it transpired that a French astronomer named Liais, who lived in Brazil, had been watching the Sun at the time of Lescarbault's famous observation, and had seen nothing unusual.

There, for some time, the matter rested; Le Verrier was fully convinced of the reality of Vulcan, but other astronomers were doubtful. Meanwhile, Le Verrier had carried out other important researches; in particular he had investigated the orbits of meteors, and in 1867 he published an excellent orbit for the Leonid stream. Unfortunately he was less fortunate in his personal relationships with other astronomers, and it seems that he was frequently ill-mannered. One of his associates commented dryly that he may not have been the most detestable man in France, but was certainly the most detested! Matters came to a head in 1870, and Le Verrier was asked to resign the Directorship of the Paris Observatory, which he did. He was succeeded by Charles Delaunay, an expert on the movements of the Moon; but in August 1872 Delaunay was drowned when his boat capsized off Cherbourg, and Le Verrier was reinstated. He remained as Director until his death, on 23 September 1877.

The story of Vulcan did not end there, and there was a flurry of excitement in 1878, when two observers – Lewis Swift and James Watson – claimed to have discovered the planet during the total eclipse on 29 July of that year. However, the observations were contradictory, and it now seems certain that Swift and Watson recorded nothing more remarkable than faint stars. Since then Vulcan has vanished – and because the irregularities in the movements of Mercury have been satisfactorily explained by Einstein's theory of relativity, it is safe to say that Vulcan never existed at all. Le Verrier was wrong.

Yet despite this mistake, and despite his irritable nature, Le Verrier accomplished much, and there is no doubt that he must be ranked as one of the greatest astronomers of the nineteenth century.

FOMALHAUT

September is a good time to locate Fomalhaut in Piscis Austrinus (or Piscis Australis), the Southern Fish. Fomalhaut is the southernmost of the first-magnitude stars to be visible from Britain or the latitude of New York; from New York or London it may attain a respectable altitude, but from North Scotland it barely rises. It is of spectral type A, and is 24 light-years away, with a luminosity 13 times that of the Sun. It may be found below the Square of Pegasus, and is easily identified because there are no bright stars anywhere near it. Piscis Austrinus is a small constellation, and apart from Fomalhaut it is entirely unremarkable.

October

Summer Time in Great Britain and Northern Ireland ends on 23 October.

New Moon: 12 October   *Full Moon:* 26 October

Mercury is in superior conjunction on 18 October, and will not be visible during the month.

Venus is still a brilliant morning star (magnitude $-3\cdot4$) rising in the east about two hours before the Sun.

Mars rises in the late evening and is in the south at sunrise. The planet moves quite rapidly through the constellation Gemini, passing 6 degrees south of Pollux on 13 October and moving into Cancer at the end of the month, when it will be seen below and in line with the Twins. The increasing brightness of Mars (magnitude $+0\cdot8$ to $+0\cdot4$) should be easily noticed. The path of Mars in the last three months of the year is shown in the diagram in the February notes (page 57).

Jupiter reaches a stationary point on 24 October on the western side of the figure of Gemini (see diagram on page 84). The planet rises in mid-evening and is growing brighter (magnitude $-1\cdot9$ to $-2\cdot1$) as it approaches opposition at the end of the year.

Saturn rises about an hour after midnight and is to be seen in Leo near the star Regulus. (Magnitudes: Saturn $+0\cdot8$, **Regulus** $+1\cdot3$).

MONTHLY NOTES NORTHERN · OCTOBER

A total eclipse of the Sun on 12 October will be visible only in the North Pacific Ocean and northern parts of South America. It will be seen as a partial eclipse in North America. For details see notes on page 126.

Two Space Anniversaries

This month the Space Age is twenty years old. It was on 4 October 1957 that the Russians launched their first artificial satellite, Sputnik 1, to the surprise of many observers in Europe and America. It must be admitted that reactions were not uniformly favourable, and one United States Army officer described the Sputnik as 'a hunk of old iron that almost anybody could launch' – though at that stage the American programme was floundering badly. Sputnik 1 carried little apart from a radio transmitter, but its famous 'Bleep! bleep!' signals will not be forgotten by anyone who heard them. The satellite did not stay aloft permanently; it re-entered the dense atmosphere, and was destroyed, early in January 1958.

Another Russian triumph took place in October 1959. This was the flight of Luna 3 (or Lunik 3), which went on a journey round the Moon and sent back the first pictures of the far side. By modern standards these pictures were blurred, and there were some strange misinterpretations; the Russians announced the discovery of a major range of peaks which they named the Soviet Mountains, subsequently found to be non-existent. However, the dark-floored walled formation now called Tsiolkovskii, in honour of the great Russian space pioneer, was clearly shown, together with various other features. Luna 3 was expected to send back more pictures during its second pass behind the Moon, but contact with it was suddenly lost, and was never regained. The Soviet authorities suggested that it might have been hit by a meteoroid, but it seems more likely that a fault developed in the transmitting apparatus. The final fate of Luna 3 is unknown.

Sun and Moon

The Solar eclipse due on 12 October will be total; the Moon will cover the Sun completely, and for a brief period the chromosphere and the corona will shine out. The eclipse on 18 April last was annular, and the Moon did not appear quite large enough to blot out the whole of the brilliant solar disk.

The fact that the Sun and the Moon appear almost exactly the same size in the sky is nothing more than coincidence – but it is a very fortunate one for astronomers. The apparent diameters have to be given in angular measure; to say that, for instance, the Sun looks about an inch across is no more useful than saying that the Sun appears the same size as a piece of wood! The values for the Sun range from $32' 35''$ when the Earth is at perihelion down to $31' 31''$ when the Earth is at aphelion; for the Moon, the limiting values are $33' 31''$ and $29' 22''$. The dark cone of shadow cast by the Moon is only just long enough to reach the Earth, and no total eclipse can last for as long as eight minutes; most are much shorter than this, and there are some total eclipses which are annular along part of the track. Moreover, the zone of totality can never be more than 170 miles broad, which explains why total eclipses are rare as seen from any particular place on Earth. From England, the last total eclipse took place in 1927; the next will not be until 11 August 1999, when the track will cross part of Cornwall. It is interesting to note the years of total eclipses visible from England: 1424, 1433, 1598, 1652, 1715, 1724 and then 1927 and 1999, after which we must wait until 7 October 2135.

Lunar eclipses are much more common from any particular site, because when an eclipse of the Moon occurs it is visible over a complete hemisphere of the Earth instead of from a relatively narrow track.

TRANSFERRED STARS!

During October evenings the Square of Pegasus is prominent in the south, and cannot be mistaken even though its stars are not outstandingly bright. The Square is made up of four stars: Alpha, Beta, and Gamma Pegasi, and Alpha Andromedæ. Details of these stars are as follows:

Star	Mag.	Spectrum	Distance light-years	Abs. mag.
Alpha Pegasi (Markab)	2·50	B9	109	−0·1
Beta Pegasi (Scheat)	var.	M2	210	−1·5
Gamma Pegasi (Algenib)	2·84	B2	570	−3·4
Alpha Andromedæ (Alpheratz)	2·06	B9	90	−0·1

Though Gamma Pegasi appears the faintest of the four, it is actually the most luminous, and seen from the standard distance of 10 parsecs it would be comparable with Venus. Beta Pegasi is an orange variable; the range is from magnitude $2\frac{1}{4}$ to $2\frac{3}{4}$, and there is a very rough period of between five and six weeks. Binoculars bring out its colour excellently, and show that it is very different from its three companions.

In view of the obvious pattern, it may be asked why one of the Square stars – Alpheratz – is included in the adjacent constellation of Andromeda. It used, in fact, to be in Pegasus, and was then known as Delta Pegasi; its transfer seems to be somewhat illogical. This is one of a few cases of stars which have been transferred from one constellation to another in comparatively modern times. Sigma Scorpii used to be included in Libra, and a more notable case is that of Beta Tauri (Alnath), which was formerly known as Gamma Aurigæ. Since the constellation boundaries have now been finally laid down by decree of the International Astronomical Union, it is unlikely that there will be any more cases of transferred stars.

THE 'GARNET STAR'

The rather obscure constellation of Cepheus is near the zenith during October evenings. Part of it lies between Polaris and the W of Cassiopeia, but there is no really distinctive shape, and the only star above the third magnitude is Alpha or Alderamin, magnitude 2·44 and with a spectrum of type A7.

Cepheus contains Delta Cephei, the prototype short-period variable. Another interesting variable is Mu Cephei, which is never conspicuous with the naked eye; at its maximum it may reach magnitude 3·7, while at minimum it may drop to almost magnitude 6. It seems to be completely irregular. It is notable because of its very red colour; Sir William Herschel nicknamed it 'the Garnet Star' and with binoculars its hue is very obvious. though with a telescope the colour is very much enhanced. The spectral type is M; Mu Cephei is a very luminous, remote giant.

Exact comparisons are difficult, but certainly Mu Cephei is one of the reddest stars to be visible with the naked eye, and for this reason it is well worth locating.

November

New Moon: 11 November   *Full Moon:* 25 November

Mercury is not likely to be seen until the last days of the month when it begins to appear as an evening star. Although quite bright before its greatest elongation from the Sun, it is very low in the south-west at sunset.

Venus is now drawing closer to the Sun, and rises less than two hours before sunrise (magnitude $-3 \cdot 4$).

Mars rises in the late evening, and will be easily recognized as it is now very bright and is in a rather barren part of the sky in Cancer. In mid-November, Mars will be near Delta Cancri and the Praesepe cluster (see diagram on page 57). The brightness of the planet increases by half a magnitude during the month

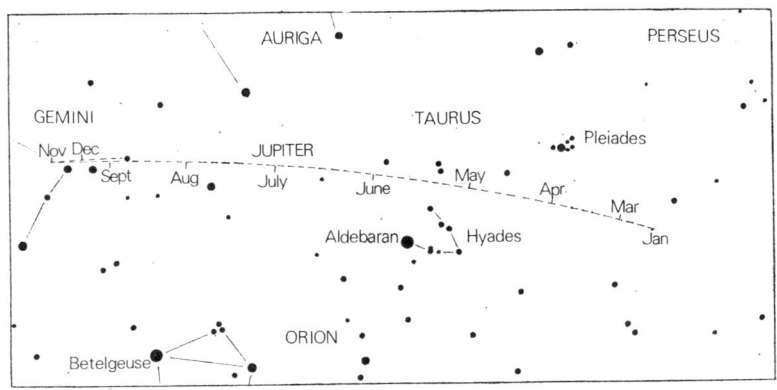

Jupiter, 1977

(+0·4 to −0·1) as its distance from the Earth decreases by about 20 million miles.

Jupiter is now approaching opposition, and is moving retrograde in Gemini, on the western side of the well-known figure. The planet rises shortly after dark and is the brightest planet in the evening sky.

Saturn is moving direct in Leo, and passes less than a degree north of Regulus on 3 November. The planet rises at midnight at the beginning of the month. The rings have now closed even more, and the planet is less bright, but is still half a magnitude brighter than Regulus. (Magnitudes: Saturn +0·8. Regulus +1·3). The darker evenings give an opportunity to look for the satellite Titan, which will be at eastern elongation on the nights of 1 and 17 November. The rings are still inclined sufficiently to allow Titan to pass clear of the planet to the north or south at conjunctions.

THE SATELLITES OF SATURN

Saturn has ten known satellites – more than for any other planet in the Solar System apart from Jupiter. However, only Titan is really large. Its diameter is rather uncertain, but it is certainly greater than that of the Moon, and Titan is the only known planetary satellite to have an appreciable atmosphere. It was described in detail by Dr Garry Hunt in the 1976 *Yearbook*.

Data for the satellites are as follows:

Name	Discoverer and year	Mean distance from centre of primary, in thousands of miles	Sidereal period d h m	Max. mag.
Janus	Dollfus, 1966	98	17 58	14
Mimas	Herschel, 1787	113	22 37	12·1

MONTHLY NOTES NORTHERN · NOVEMBER

Name	Discoverer and year	Mean distance from centre of primary, in thousands of miles	Sidereal period			Max. mag.
			d	h	m	
Enceladus	Herschel, 1787	149	1	8	53	11·6
Tethys	Cassini, 1684	183	1	21	18	10·6
Dione	Cassini, 1684	235	2	17	41	10·7
Rhea	Cassini, 1672	328	4	12	25	9·7
Titan	Huygens, 1655	760	15	22	41	8·2
Hyperion	Bond, 1848	920	21	6	38	13
Iapetus	Cassini, 1671	2200	79	7	56	9
Phœbe	Pickering, 1898	8050	550	10	50	14

An extra satellite was reported by Pickering in 1904. It was said to move in an orbit between those of Titan and Hyperion, and it was even given a name: Themis. However, it has never been seen again, and there now seems little doubt that Pickering mistook a faint star for a satellite.

Iapetus is of special interest, because it is very variable in brilliancy. When west of Saturn it can become an easy telescopic object, and some recent estimates make its maximum magnitude 9; but when east of Saturn, it becomes faint – well below magnitude 11. Either it is irregular in shape, or (more probably) it has its two hemispheres of unequal reflecting power. It seems definite that Iapetus, like all large planetary satellites, keeps the same face turned towards its primary, which explains why the fluctuations in brightness are regular.

Dione is also of interest, because it seems to be much more massive than Tethys even though it is no brighter. Indeed, Dione is probably the densest of all the satellites of Saturn. A modest telescope will show it, together with Rhea and Tethys. Mimas, Enceladus, and Hyperion are much more elusive; Hyperion because of its faintness and the other two because they are inconveniently close to Saturn. Phœbe is an oddity; not only is it very remote from Saturn, but it moves in a retrograde orbit. It may well be a captured asteroid rather than a true satellite.

Janus, the latest addition to the Saturnian family, is very close to the planet, and is observable only when the rings are edgewise-on to us, so that there is no chance of locating it in 1977. Its existence is doubted by some astronomers, and every effort will be made to confirm it when the rings are edge-on again in 1980. Of course, there may be other minor satellites awaiting discovery.

The November Leonids

The Leonid meteors may at times provide spectacular showers. Formerly, brilliant displays occurred about once in 33 years; thus there were major showers in 1799, 1833, and 1866, but the meteor stream was then perturbed by the gravitational pulls of the planets, and the expected showers of 1899 and 1933 did not materialize. In 1966 the Leonids were back, and a magnificent display was seen in some parts of the world (including the United States), though the shower took place during daytime in Europe.

It does not seem likely that there will be a major shower in 1977, but a few Leonids will doubtless be seen, and it is worth keeping a watch on the night of 17 November. The Moon will be six days old, and will not interfere with meteor-watching to any great extent.

The W of Cassiopeia

Apart from Ursa Major (the Great Bear), Cassiopeia is probably the most famous of the constellations of the far north. It is circumpolar in Britain, and is almost overhead during November evenings. Its five main stars are arranged in the form of a W, and are very easy to identify. The five are Alpha or Shedir (magnitude 2·3 – probably slightly variable), Beta (2·3), Gamma (variable), Delta (2·7), and Epsilon (3·4).

Gamma Cassiopieæ is the most interesting of the five. It is an unstable star, and is definitely variable. In 1936 the magnitude rose to 1·6, but for years now the average value has been about 2·3. It is worth observing, particularly since it can be followed with the naked eye; Beta makes a good comparison. Alpha Cassiopeiæ is of type K, and is definitely reddish. It has long been classed as a variable of small range, though in a well-known catalogue by the Russian astronomer, Kukarkin, it is given as being constant in brightness.

Cassiopeia lies in the Milky Way, and the whole region is very rich. It was here, too, that Tycho's supernova blazed out in 1572, becoming bright enough to be seen with the naked eye in broad daylight; it is still identifiable as a radio source.

KARL VON LITTROW

November 16 is the centenary of the death of Karl Ludwig von Littrow, former Director of the Vienna Observatory. His father, Johann von Littrow (1781-1840), was also an astronomer, who held appointments at Cracow, Kasan, and Budapest before becoming Director of the Vienna Observatory in 1819; he was one of the first astronomers to recognize the existence of the Sun's chromosphere, and was also the author of various popular books.

Karl was born on 18 July 1811, and in 1831 became assistant to his father at Vienna, becoming Director in 1842. He is best remembered for his work in connection with the asteroids; he made no actual discoveries, but he computed the orbits of many of these interesting little members of the Solar System.

December

New Moon: 10 December   *Full Moon:* 25 December

Solstice: 21 December

Mercury is at greatest eastern elongation (21°) on 3 December and is then an evening star, but is very low in the south-west and not at all conspicuous. Mercury is in inferior conjunction on 21 December.

Venus now rises only about an hour before the Sun, and by the end of the year will no longer be visible as a morning star. Superior conjunction takes place in January of next year.

Mars is at a stationary point on 13 December in Cancer, to the west of the sickle of Leo, and rises in mid-evening. Mars is easy to recognize, since it is now a really brilliant object (magnitude −0·2 to −0·7). The planet will be nearly half a magnitude brighter than this when it reaches opposition in January of next year.

Jupiter is at opposition on 23 December and is moving retrograde in Gemini. By the end of the year it will be seen on the borders of Gemini and Taurus. The distance of Jupiter from the Earth at opposition is 386 million miles (621 million km) which is about 12 million miles greater than the distance at the 1976 opposition. As a result Jupiter is slightly less brilliant (magnitude −2·3). The next opposition of this planet occurs in January 1979.

Saturn is at a stationary point on 12 December near the star Regulus in Leo. After this the planet begins to move retrograde and grows brighter as it moves towards opposition in February 1978 (magnitude +0·7 to +0·6). By the end of the month, Saturn rises in mid-evening to the north of east. The satellite Titan may be seen at eastern elongation on the nights of 3 and 19 December.

THE SATELLITES OF JUPITER

The discovery of a possible fourteenth satellite of Jupiter was made in September 1975 by Charles Kowal, just a year after his discovery of the thirteenth satellite. Both discoveries were made with the 48-inch Schmidt telescope at Mount Palomar, but although J XIII has been well observed and its orbit determined, there are as yet no details concerning J XIV. Another announcement of some interest was made by a working party of the International Astronomical Union, recommending names for all of the Jovian satellites. Although the five inner satellites have had classical names (like satellites of all the other planets), these are not often used, and Jupiter's moons have been referred to by Roman numerals which indicate the order of their discovery. This system has many disadvantages, though whether the new names will be easier to remember is a matter of opinion.

The names of the thirteen satellites are given here in order of their distance from Jupiter:

No.	Name	Period (days)	Discovered	
V	Amalthea	0·50	Barnard, 1892.	Lick
I	Io	1·77	Galileo	
II	Europa	3·55	Galileo	1610. Padua
III	Ganymede	7·15	Galileo	
IV	Callisto	16·69	Galileo	
XIII	Leda	238	Kowal, 1974.	Mt Palomar
VI	Himalia	251	Perrine, 1904	Lick
X	Lysithea	259	Nicholson, 1938	Mt Wilson
VII	Elara	260	Perrine, 1905	Lick

No.	Name	Period (days)	Discovered	
XII	Ananke	631	Nicholson, 1951	Mt Wilson
XI	Carme	692	Nicholson, 1938	Mt Wilson
VIII	Pasiphaë	744	Melotte, 1908	Greenwich
IX	Sinope	758	Nicholson, 1914	Lick

The satellites divide themselves into three distinct groups. The inner five (the only ones to be discovered visually) travel in nearly circular orbits in the plane of Jupiter's equator. The four big satellites discovered by Galileo form a tightly bound system, perturbing each other in such a way that it is impossible to deal with the motion of any one of them without including the other three. These four are quite easy to see with any sort of glass, but the others are all very small and faint, and are only photographed with the largest telescopes.

The second group contains four satellites which revolve in the normal direct sense at a distance of some seven million miles from Jupiter. Their orbits are eccentric and inclined at about 28° to the plane of Jupiter's equator, but they are interlocked and are inclined at about 28° to each other, so that there is never any possibility of collision. They are severely perturbed by the attraction of the Sun, so that the distances and periods can vary considerably.

The satellites in the outermost group, revolving at about 14 million miles from Jupiter, travel in the retrograde sense. This has the effect of reducing the time during which the very severe solar perturbations can act. As a result these orbits are quite stable, but they change so much that it is impossible to give any precise figures for their orbits; those given above are published mean values. The names for satellites VI to XIII follow the rule that those orbits with direct motion have names ending in *a*, while those with retrograde motion have names ending in *e*.

WHEN MARS DRAWS NEAR

Mars is now approaching opposition once more, and has be-

MONTHLY NOTES NORTHERN · DECEMBER

come very prominent. The actual date of opposition is 22 January 1978. However, this is not a very favourable apparition, since Mars is not far from its aphelion or furthest distance from the Sun. The distance from Earth will never be less than 60,000,000 miles, as against less than 35,000,000 miles at the opposition of 1971. It may be of interest to give details of the oppositions of Mars between 1970 and 1980:

Year	Date	Closest approach to Earth, miles	Maximum angular diameter, secs. of arc
1971	10 Aug.	34,920,000	24·9
1973	25 Oct.	40,530,000	21·4
1975	15 Dec.	52,570,000	16·2
1978	22 Jan.	60,720,000	14·3
1980	25 Feb.	62,960,000	13·8

In early 1978 the small apparent diameter of Mars will be partly counteracted – for northern observers – by the planet's position. Mars will be in Cancer, well north of the celestial equator. However, telescopes of some size will be needed to make any worth-while observations. An 8-inch reflector will show the main features, including the dark areas, but an aperture of over 12 inches is required to show the finer detail. At opposition Mars will be of magnitude $-1\cdot1$, so that it will not quite be the equal of Sirius.

THE PLEIADES AND THE HYADES

Taurus, the Bull, has now come back into the evening sky. It contains many interesting objects, including the famous Crab Nebula (Messier 1), but it is probably most famous because of the presence of two open star clusters, the Pleiades and the Hyades-

The two clusters are very different. The Pleiades are recognize able at a glance, and it is interesting to see how many stars can b, counted without optical aid; the average number, on a clear night is seven (hence the nickname of 'the Seven Sisters'), but keen-eyed people can see more, and it is said that a last-century German astronomer, E. Heis, could see 19. Photographs show that the stars of the Pleiades are contained in nebulosity. The leader of the group is Alcyone or Eta Tauri, of the third magnitude; a hundred years ago Johann Mädler, the great observer of the Moon, was maintaining that Alcyone was the central star of the Galaxy – a conclusion long since found to be baseless.

The Hyades are more scattered, and are rather overpowered by the strong orange light of Aldebaran. In fact Aldebaran is not a member of the cluster, and lies about half-way between the Hyades and ourselves, so that it appears 'in the foreground', so to speak. The Hyades are best seen with low-power binoculars, and are much less rich than the Pleiades. One of the Hyades stars, Theta Tauri, is a naked-eye double.

MONTHLY NOTES, 1977
SOUTHERN HEMISPHERE

In this edition of the *Yearbook*, full monthly notes for Southern Hemisphere observers are given for the first time. Of course, most of the familiar constellations from the viewpoint of European observers are also to be seen in countries such as South Africa and Australia, the only really notable exceptions being the two Bears and Cassiopeia.

The notes which follow are for the approximate latitude of 26 degrees south, and so are applicable to South Africa, most of Australia, and much of South America. Some adjustment must be made for New Zealand, which is appreciably further south, and for Rhodesia, which is closer to the equator.

The far southern constellations are given in four separate charts. Those who live in the North must remember that Southern Hemisphere observers see a reverse view; thus in Orion Rigel is higher than Betelgeux, and the Belt stars point upward to Sirius and downward to Aldebaran! Brief notes on the planets are given where appropriate; thus in the early part of 1977 Saturn is in the Cancer-Leo area, high up from London or New York but rather inconveniently low from Sydney or Johannesburg.

January

During January evenings the brilliant star Achernar, in Eridanus, is almost overhead, and cannot be mistaken; the Southern Cross is almost at its lowest. The south polar area of the sky, shown opposite, is extremely barren. The pole itself lies roughly midway between Achernar and the Cross, but there is no bright star anywhere near it. The south polar star, Sigma Octantis, is only of the fifth magnitude. However, the two Clouds of Magellan are high up and are very prominent. The Large Cloud remains visible with the naked eye even during moonlight.

Orion is very prominent, together with Sirius in Canis Major and Canopus in Carina (see the diagram on page 113). Fomalhaut is to be seen in the south-west, and the Square of Pegasus lies above the western horizon.

THE PLANETS

Mercury and **Mars** are badly placed. **Venus** is an evening star, but is not yet at its best.

Jupiter, on the boundaries of Aries and Taurus, attains a respectable altitude and is very conspicuous in the north. **Saturn**, on the borders of Cancer and Leo, rises in the late evening. However, neither of the two giant planets is as well placed from South Africa or Australia as from Europe or the United States.

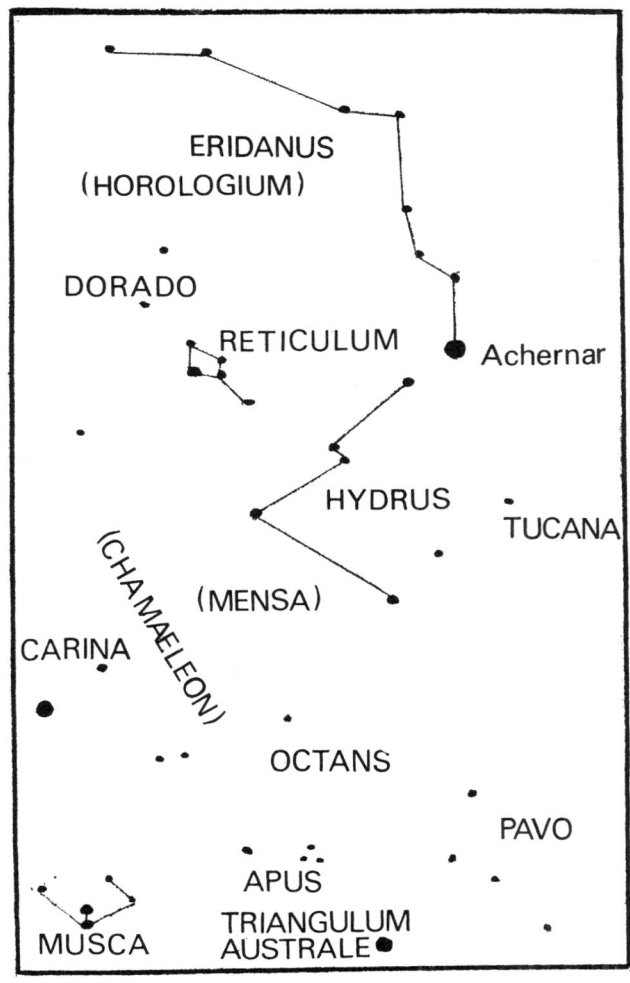

February

Orion is now at its best during the evening, and dominates the northern aspect of the sky; Capella is visible, very low down, and both Sirius and Canopus are high. It is interesting to compare the two; Sirius is almost a magnitude the brighter, though it is, of course, one of our nearest stellar neighbours, whereas Canopus is extremely luminous and remote. Crux is coming into prominence in the south-east together with the two brilliant Pointers, Alpha and Beta Centauri; Fomalhaut is decending in the south-west, and the Twins, Castor and Pollux, are rising in the north-east. Much of the eastern aspect is occupied by the vast, barren constellation of Hydra. Now that Argo Navis has been divided up, Hydra is the largest separate constellation in the sky (it is worth noting that Crux Australis, the Southern Cross, is the smallest!) Achernar remains prominent, and now that Crux is higher up it is easier to locate the position of the south celestial pole.

THE PLANETS

Jupiter remains prominent in the north; **Saturn** rises in mid-evening, but is rather low down. **Venus** is still an evening star, but sets rather earlier than in January, while **Mars** is out of view.

March

March evenings provide an ideal time for identifying Canopus, which is almost at the zenith. Locating it is an easy task, because it is much brighter than any other star apart from Sirius. It has an F-type spectrum, and is often said to be yellowish, but most observers will call it pure white. It is the leader of Carina, the Keel of the old Ship (Argo), which is an extremely rich area, well worth sweeping with binoculars. The False Cross lies partly in Carina and partly in Vela (the Sails of the Ship); its form is very like that of Crux, but it is rather larger, and its stars are not so brilliant, though they are of about the second magnitude. The Ship is shown in the diagram given here. Broadly speaking, it lies between Sirius on the one side and Crux on the other.

The Clouds of Magellan remain high up; Achernar is now in

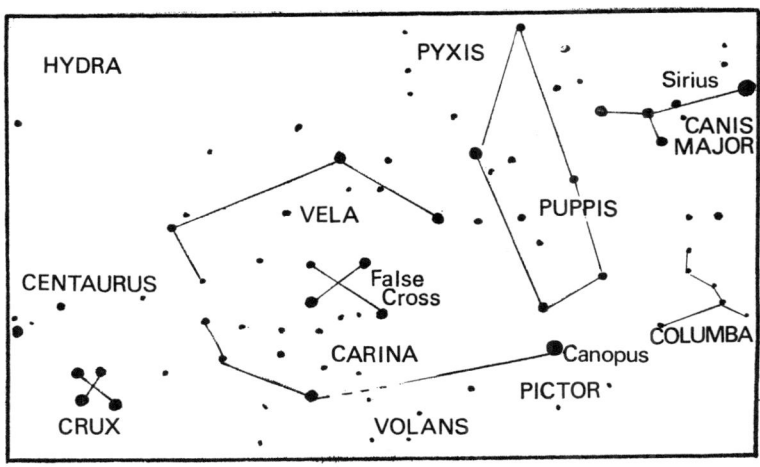

the south-west and Crux in the south-east. Orion is still very prominent, with Gemini reasonably high in the north and Leo rising in the north-east. Fomalhaut has practically set in the south-west.

THE PLANETS

Mercury is out of view for most of the month, and even at the end of March it will not be seen with the naked eye, as it lies well north of the celestial equator. **Venus** sets not long after the Sun, and **Mars** will hardly be seen. **Jupiter** now sets in mid-evening. **Saturn**, in Cancer, is high enough to be reasonably well observed in the late evening.

April

Crux is now high up, with Achernar starting to descend in the south-west; the Clouds of Magellan remain well placed. This is a good time for viewing the Milky Way, which runs from the south-eastern horizon through Crux and Argo through to the Orion area and Auriga; Capella has virtually set in the north-west. Orion itself is prominent, still quite high in the west, with Sirius even higher and Canopus not far from the zenith. Fomalhaut has set.

Hydra occupies much of the eastern aspect, and Leo has now risen. Late in the evening part of Ursa Major can be seen. The main 'Plough' is never well seen from South Africa or most of Australia, and never rises over New Zealand, but from Rhodesia it can be quite prominent, though it is always low.

THE PLANETS

Mercury is at elongation on 10 April, but this is not a favourable apparition for Southern Hemisphere observers. **Venus** passes through inferior conjunction early in the month; **Jupiter** sets very early, but **Saturn** is still quite prominent in the north. **Mars** is still badly placed, and is still very remote; it is about as bright as Beta Crucis, the second star of the Southern Cross, but rises only shortly before the Sun.

The annular eclipse of the Sun on 18 April will be visible from parts of Africa. See the notes on page 126.

May

Orion now sets very early; Sirius remains visible in the south-west and Canopus is still prominent. Crux is nearing the zenith, and makes as plendid group with Centaurus (see the diagram on page 117.) Achernar is low in the south, and in the north Ursa Major may be seen in part. In the east, Scorpius is coming into prominence; Virgo is high, and Leo is at its best. In the north-east the brilliant orange Arcturus has come into view. Arcturus is actually the brightest star in the sky apart from Sirius, Canopus, and Alpha Centauri, but it is well north of the celestial equator and is therefore never seen to best advantage.

Virgo is now prominent, and near it, very high up, is the rather distinctive quadrilateral of stars marking Corvus. This is also a good time to look for Ursa Major, very low in the north.

THE PLANETS

Mercury and **Mars** are very badly placed, and **Jupiter** is now out of view. **Venus** is at maximum brightness on May 11. **Saturn** is still visible until mid-evening, but sets before midnight.

June

Crux this month is very high during the evening, and together with Centaurus it makes up a splendid group. Moreover, the Milky Way runs through it. The area is shown in the diagram, together with Scorpius or Scorpio, the Scorpion – one of the few constellations which really does give some impression of the object it is meant to represent. Much of Scorpius, including Antares may be seen from Europe, but Southern Hemisphere observers alone can see the constellation at its glorious best. Near the 'sting' there are some magnificent open clusters, and many rich star fields. The region between the Scorpion and the Centaur is occupied by Lupus, the Wolf, which has several reasonably bright stars but lacks any distinctive shape.

The Pointers, Alpha and Beta Centauri, are truly splendid.

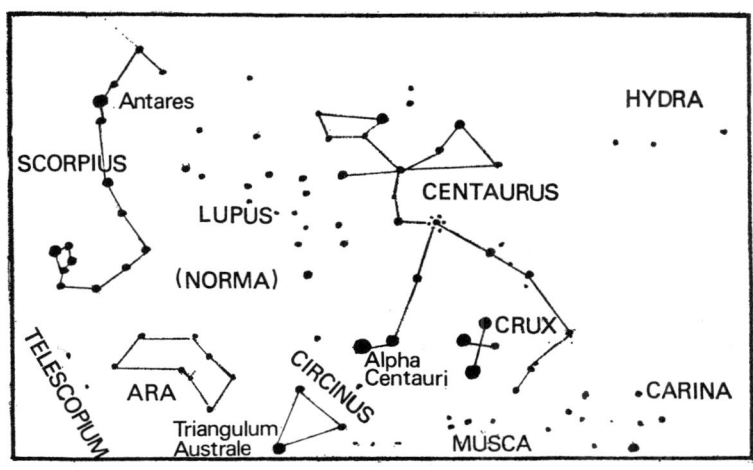

Adjoining them is the Southern Triangle, Triangulum Australe; the main star, Alpha or Atria, is above the second magnitude, and is decidedly orange. Between Triangulum Australe and the Scorpian is the reasonably well-marked group of Ara, the Altar.

THE PLANETS

This is a poor month for the planets. **Mercury** and **Jupiter** are out of view; **Venus** and **Mars** rise in the early hours; **Saturn** sets in mid-evening. **Neptune** not far from Antares, is well placed, but is of course much too faint to be seen with the naked eye.

July

Crux and Centaurus are still high, slightly west of the zenith in mid-evening. Scorpius is superb in the south-east; the red Antares dominates the group, but there are several other bright stars in the chain. Sirius has now gone, and Canopus and Achernar are not far above the horizon. This is the best time of the year for observing Arcturus, which attains a reasonable altitude in the north; note too the prominent little semi-circle of stars marking Corona Borealis, the Northern Crown. The north-eastern aspect is largely occupied by Hercules, which lacks any brilliant stars, and Ophiuchus; Altair in Aquila is rising in the east, and the bright but rather ill-formed constellation of Sagittarius is prominent. This is one of the very richest areas of the Milky Way, and is worth sweeping with binoculars. To the west Virgo is still prominent, but Leo is disappearing below the horizon.

THE PLANETS

Again this is a poor month for southern observers, with **Mercury** and **Saturn** out of view and **Venus**, **Mars**, and **Jupiter** all very badly placed in the morning sky.

August

Scorpius is now at the zenith, and is at its very best; so too is the Milky Way, running from Aquila in the east through Sagittarius, Scorpius, and Crux down to the south-western horizon. Crux and Centaurus are still high and prominent, but Canopus is at its lowest, and is barely above the horizon during evenings this month. Fomalhaut has come back into view in the south-east, and Virgo is still visible in the west.

Arcturus remains above the horizon, but is now low in the north-west. Vega can be seen in the north-east, low down; the northern aspect is largely filled by Hercules, Ophiuchus, and Serpens, so that it appears rather barren. Under favourable conditions a few stars of Draco may be seen just above the northern horizon.

THE PLANETS

Mercury is an evening star, but is not at all prominent; it may be glimpsed early in the month. **Venus** in Gemini, rises in the dawn light. **Mars** comes up rather earlier, but is still no brighter than Alpha Crucis, and is still moving north in the sky; by the end of the month it is close to **Jupiter**, but both planets are inconveniently low. **Saturn** is still out of view.

September

Crux is descending in the south-west, Achernar rising in the south-east; by mid-evening the two are at roughly the same altitude, so that the south celestial pole is easy to locate. Fomalhaut has become high in the south-east; Sagittarius is overhead, Scorpius not far to the west of the zenith. Spica is setting in the west.

This is a good time to see the bright stars which make up what Northern Hemisphere observers call the Summer Triangle – a term which certainly does not apply to dwellers in the south. Altair, in Aquila, is the highest of the three, and is easy to locate in the north; Vega is lower and to the west; Deneb lower still, not far above the north-eastern horizon. The cross of Cygnus may be made out, but is never seen to advantage. The Square of Pegasus rises in the east later in the evening; Hercules is descending in the north-west.

THE PLANETS

Mercury is an evening star, but well north of the celestial equator, so that to southern observers this is not a favourable elongation. The bright planets are still badly placed; **Venus** and **Saturn** are in the region of Leo, and rise not long before the Sun, though **Mars** and **Jupiter**, both in Gemini, rise well before dawn.

October

During October evenings the Southern Cross, the Centaur, and Canopus are all very low, and though Scorpius is still prominent it is sinking in the south-west. Achernar is high in the south-east, and the Square of Pegasus has come into view in the north-east; the triangle made up of Altair, Vega, and Deneb remains visible in the north-west.

Fomalhaut is very high, and the area near the zenith is occupied by the so-called Southern Birds: Grus (the Crane), Pavo (the Peacock), Phœnix (the Phœnix) and Tucana (the Toucan). This is one of the more confusing regions of the sky but at least Grus is easy to find; it has a distinctive shape, and both its leading stars (Alnair and Beta Gruis) are of the second magnitude. Their colours are different, since Alnair is white and Beta a lovely warm

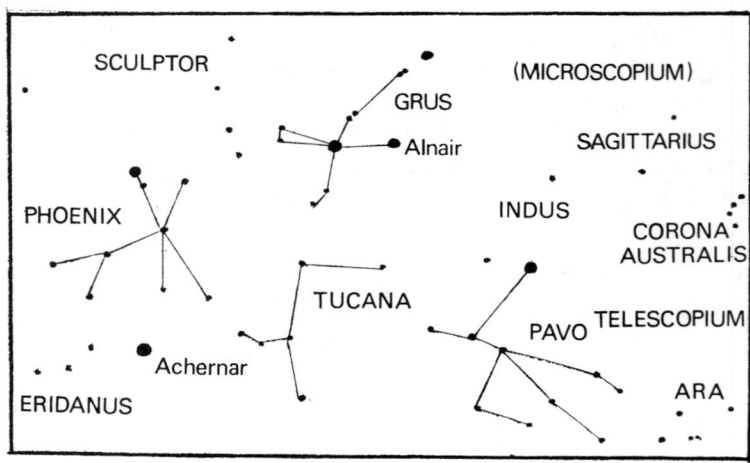

orange. Tucana is a faint group, but it contains the globular cluster 47 Tucanæ – which, with the exception of Omega Centauri is the finest in the sky, and lies in the same general area as the Clouds of Magellan. All four Birds are shown in the diagram given here, together with Achernar in Eridanus.

THE PLANETS

Mercury is out of view; **Venus** is badly placed as a morning star. By now **Mars** and **Jupiter**, in Gemini, rise some hours before the Sun, though Mars is still travelling quickly and passes into Cancer before the end of the month. **Saturn**, in Leo, will scarcely be seen.

The total Solar eclipse of 12 October will be seen from some parts of South America. See the notes on page 126.

November

The Southern Cross is at its very lowest during November evenings, and the Scorpion is setting in the south-west; Canopus is gaining altitude in the south-east, but is not yet very high. In the north-west, Vega and Deneb have virtually disappeared, and Altair is extremely low, setting well before midnight. Fomalhaut is at the zenith, and the four Southern Birds remain high; so too is Achernar, and this is also a good time for studying the Clouds of Magellan, together with the globular cluster 47 Tucanæ.

The Square of Pegasus is on view in the north, and is quite prominent. Leading off from it are the stars of Andromeda, but the altitude is so low that from South Africa and most of Australia it is by no means easy to locate the Great Spiral, Messier 31, which is so familiar to northern observers.

THE PLANETS

Mercury is an evening star at the end of the month, but is not prominent; **Venus** is disappearing in the morning dawn. **Jupiter** (in Gemini), **Mars** (in Cancer), and **Saturn** (in Leo) are all morning objects, rising in the early hours; **Jupiter** has become really prominent, while **Mars** is now almost as brilliant as Alpha Centauri, and the conjunction of **Saturn** with Regulus on 3 November should be easily observable before sunrise.

December

Crux is still very low; it follows that Achernar, on the opposite side of the pole, is practically at the zenith. Scorpius and Sagittarius set very early in the evening; Fomalhaut remains high, and the Square of Pegasus is still prominent in the north-west before midnight. Orion now rises early on, and both Sirius and Canopus become very conspicuous by late evening. Cetus, the Whale or Sea-monster, is very high up, but contains few bright stars, and is notable chiefly because of the presence of Mira (Omicron Ceti), the prototype long-period variable.

The Clouds of Magellan are excellently placed, near the overhead point, but this is not a good time for seeing the Milky Way at its best.

THE PLANETS

Mercury may be visible as an evening star early in the month, but **Venus** has virtually disappeared into the dawn light. **Mars**, **Jupiter**, and **Saturn** are all becoming much better placed, and are prominent after midnight. By the end of the year **Mars** has become as bright as Canopus, while **Jupiter** is at opposition on 23 December and is therefore due north at midnight. Though it is relatively low, its great brilliance makes it unmistakable.

Eclipses in 1977

In 1977 there will be two eclipses of the Sun and one of the Moon; there will also be a penumbral eclipse of the Moon.

(1) *A partial eclipse of the Moon* on 4 April, visible in western Europe (including the British Isles) and in North and South America. The eclipse begins at $3^h\ 30^m$ and by $4^h\ 18^m$ about 20 per cent of the northern part of the Moon will be covered by the Earth's shadow. The eclipse ends at $5^h 06^m$, about 20^m before sunrise in southern England.

(2) *An annular eclipse of the Sun* on 18 April, the central path entering the west coast of Africa and crossing Zambia and Tanzania to end in the Indian Ocean. A partial eclipse will be visible over South and East Africa. At Dar-es-Salaam the eclipse begins at $9^h\ 11^m$ and will be annular from $11^h\ 12^{m}\cdot 3$ until $11^h\ 18^{m}\cdot 8$. The eclipse is also visible in the Seychelles, beginning at $10^h\ 11^m$, becoming annular from $11^h\ 55^{m}\cdot 0$ until $12^h\ 00^{m}\cdot 5$.

(3) *A penumbral eclipse of the Moon* on 27 September, is visible (normally only with special equipment) in North America. The eclipse begins at $6^h\ 18^m$ and reaches a maximum at $8^h\ 29^m$ when about 93 per cent of the Moon will be in the penumbral shadow of the Earth.

(4) *A total eclipse of the Sun* on 12 October has a central path extending from the North Pacific Ocean to the coast of South America, ending in Venezuela. The eclipse is nearly total at Bogotá at sunset, and will be seen as a partial eclipse in most of North America.

Occultations in 1977

In the course of its journey round the sky each month, the Moon passes in front of all the stars in its path and the timing of these occultations is useful in fixing the position and motion of the Moon. The Moon's orbit is tilted at more than five degrees to the ecliptic, but it is not fixed in space. It twists steadily westwards at a rate of about twenty degrees a year, a complete revolution taking 18·6 years, during which time all the stars that lie within about six and a half degrees of the ecliptic will be occulted. The occultations of any one star continue month after month until the Moon's path has twisted away from the star but only a few of these occultations will be visible at any one place in hours of darkness.

Only four first-magnitude stars are near enough to the ecliptic to be occulted by the Moon; these are Regulus, Aldebaran, Spica, and Antares. None of these stars will be occulted in 1977, the Moon's orbit having twisted round so that the series of occultations of Spica (March 1975 to December 1976) can no longer occur, the Moon passing north of Spica each month. This change in the Moon's orbit continues, and the star Aldebaran will begin to be occulted in 1978.

The planets, which have a motion of their own, do not give rise to such lengthy series of occultations, but in 1977 the slow motion of Uranus allows it to be occulted each month until July. The occultation of Uranus at midnight on 27 June will be visible in western Europe and in North America. Mercury, Venus, Mars, and Jupiter are also eclipsed in this way, but none of these events will be readily observable in the Northern Hemisphere.

Details of occultations of all stars brighter than magnitude 7·5 are given annually the *Handbook of the British Astronomical Association* (for Great Britain, Australia, and New Zealand) and in the special supplement issued by *Sky and Telescope* for observers in the United States and Canada. This special supplement is available without charge from the Director, Nautical Almanac Office, U.S. Naval Observatory, Washington, D.C., 20390.

Comets in 1977

The appearance of a bright comet is a rare event which can never be predicted in advance, because this class of object travels round the Sun in an enormous orbit with a period which may well be many thousands of years. There are therefore no records of the previous appearances of these bodies, and we are unable to follow their wanderings through space.

Comets of short period, on the other hand, return at regular intervals, and attract a good deal of attention from astronomers. Unfortunately they are all faint objects, and are recovered and followed by photographic methods using large telescopes. Most of these short-period comets travel in orbits of small inclination which reach out to the orbit of Jupiter, and it is this planet which is mainly responsible for the severe perturbations which many of these comets undergo. Unlike the planets, comets may be seen in any part of the sky, but since their distances from the Earth are similar to those of the planets their apparent movements in the sky are also somewhat similar, and some of them may be followed for long periods of time.

The number of comets under observation in any one year is much greater than is generally supposed. The following table compares the numbers of newly discovered comets, successfully predicted returns of periodic comets and comets being followed from previous years:

	1971	1972	1973	1974
New discoveries	1	6	9	5
Predicted and recovered	5	6	6	4
Still under observation	14	10	13	14
Totals	20	22	28	23

Data for 1975 are not yet complete, but 13 new comets were discovered, six of which proved to be periodic, and 4 comets were successfully recovered as a result of predictions. In 1977 the following periodic comets are expected to return to perihelion:

Comet Johnson was discovered in 1949 and has been seen at each return, the last being in 1970. The period of this comet is 6·8 years and its orbit has only a moderate eccentricity (0·39) so that even at perihelion it is still well outside the orbit of Mars.

Comet Taylor 1916 I was discovered in 1915 and has not been seen since then. It is interesting because the nucleus was seen to divide in 1916 and new predictions have suggested that there may now be two distinct comets travelling in similar orbits (period nearly 7 years) but with one comet eleven days ahead of the other.

Comet Faye is one of the more regular short-period comets. It was discovered in 1843 and had made sixteen appearances up to its last return in 1969. The orbit changes very little and has a period of 7·4 years.

Comet Kopff is another well-known comet. Discovered in 1906 it made its tenth return in 1970. The orbit has a period of 6·4 years and an inclination of less than 5 degrees.

Comet Grigg-Skjellerup was first seen in 1902, but rediscovered in 1922, and it therefore carries the names of both of the astronomers concerned. This comet has the short period of only 5·1 years, and it made its twelfth appearance in 1972. The perihelion is quite close to the Earth's orbit, and the 1977 apparition will be particularly favourable, the comet being only about 0·18 astronomical unit (less than 17 million miles) from the Earth in April 1977.

Comet Encke has the shortest known period (3·3 years) of any comet. It was first seen by Méchain in 1785 but it carries the name of the great mathematician Encke because he not only showed that the comets of 1785, 1795, 1805, and 1819 were returns of the same comet, but he also predicted its return in 1822 and this was duly observed. This was only the second prediction of the return of a periodic comet, the first being Halley's prediction of the return of the Comet Halley in 1759. Although the orbit of Encke's comet is very eccentric it is nowadays detected at each opposition in the world's largest telescopes. The return to perihelion in 1977 will be its 51st, but on this occasion the comet is poorly placed.

Comet du Toit-Neujmin-Delporte was discovered in 1941 and was not seen again until 1970. This comet has its aphelion very close to the orbit of Jupiter, so that perturbations are severe. The original period of 5·5 years has been increased to 6·3 years, the orbit being made larger while the eccentricity and inclination have diminished.

Comet Schwassmann-Wachmann (1) has a nearly circular orbit with a period of sixteen years, which lies entirely between the orbits of Jupiter and Saturn. The motion of this comet is very much like that of a planet, and it is visible each year. It is remarkable for its sudden outbursts of brightness, which seem to have some connection with solar activity.

Meteors in 1977

Meteors ('shooting stars') may be seen on any clear moonless night, but on certain nights of the year their number increases noticeably. This occurs when the Earth chances to intersect the orbit of a meteor swarm, which is a concentration of meteoric dust moving in an orbit around the Sun. Such an intersection can occur only at one particular time of year, but if the dust is spread out along the orbit, the resulting shower of meteors may last for several days. The word 'shower' must not be misinterpreted – only on very rare occasions have the meteors been so numerous as to resemble snowflakes falling.

The naked-eye study of meteors is quite a laborious task, but even a casual observer, watching for, say, ten minutes on an August night, may observe a number of Perseids. If their tracks are marked on a star map, and traced backwards, a number of them will be found to intersect in a point (or a small area of the sky) which marks the radiant of the shower. This gives the direction from which the meteors have come.

The following table gives some of the more easily observed showers with their radiants; the effect of moonlight in 1977 is indicated.

Limiting dates	Shower	Maximum	R.A.	Dec.	
Jan. 1-6	Quadrantids	Jan. 4	$15^h\ 28^m$	$+50°$	M
April 20-22	Lyrids	April 21	$18^h\ 08^m$	$+32°$	
July 27-Aug. 17	Perseids	Aug. 12	$3^h\ 04^m$	$+58°$	
Oct. 15-25	Orionids	Oct. 20	$6^h\ 24^m$	$+15°$	
Oct. 26-Nov. 16	Taurids	Nov. 8	$3^h\ 44^m$	$+14°$	
Nov. 15-19	Leonids	Nov. 17	$10^h\ 08^m$	$+22°$	
Dec. 9-14	Geminids	Dec. 14	$7^h\ 28^m$	$+32°$	
Dec. 17-24	Ursids	Dec. 22	$14^h\ 28^m$	$+76°$	M

M = moonlight interferes

Minor Planets in 1977

Although many thousands of minor planets (asteroids) are known to exist, only about 2,000 of these have well-determined orbits and are listed in the catalogues. Most of these orbits lie entirely between those of Mars and Jupiter. All of these bodies are quite small, and even the largest can only be a few hundred miles in diameter. Thus they are necessarily faint objects, and although a number of them are within the reach of a small telescope few of them ever reach any considerable brightness. Of these, the most important are the 'big four', Ceres, Pallas, Juno, and Vesta. Vesta can occasionally be seen with the naked eye, and this is most likely to occur at a June opposition, when Vesta is at perihelion. This will next happen in 1978, when Vesta reaches magnitude +5·5. In 1977, however, Vesta is no brighter than magnitude +6·6 at opposition in 9 January. The path of the planet at this time is shown in the diagram on page 57, and with a pair of binoculars it should be possible to identify the planet because of its rapid motion (about a quarter of a degree a day) towards the third magnitude star Epsilon Geminorum.

The largest of the minor planets, and the first to be discovered is Ceres, which comes to opposition on 24 March in the borders of Virgo and Coma Berenices. Although it reaches magnitude +6·5 at this time, it may be more difficult to identify in this region of the sky, where there are many faint stars, but few bright ones. The planet Juno is at opposition on 13 May on the borders of Libra and Serpens, but does not exceed magnitude +10. Pallas also comes to opposition this year, reaching magnitude +6·7 on 10 February. This planet has an orbit with the high inclination of 35°, so that it can sometimes be found well away from the ecliptic. In 1977 Pallas is at opposition low down in the constellation Pyxis, south of the head of Hydra (see star-chart 3R).

Although most of the minor planets cluster in the region between Mars and Jupiter, there are notable exceptions. The largest orbit is that of 944 Hidalgo, discovered by Baade at Bergedorf in 1920. The perihelion of the orbit of Hidalgo lies just beyond the orbit of Mars, while at aphelion it goes out to the orbit of Saturn. In contrast, the smallest orbit in the catalogue is that of 1566 Icarus, also found by Baade at Mt Palomar in 1949. This planet has its aphelion just beyond the orbit of Mars, but at perihelion it passes well inside the orbit of Mercury. An even smaller orbit now seems to have been calculated for an object discovered by Helin at Mt Palomar in the first few days of 1976. This new planet, known provisionally as 1976 AA, is unique in having a mean distance from the Sun which is less than that of the Earth. It travels in an orbit with a period of 0·949 year, its perihelion lying just inside the Earth's orbit, and its aphelion just outside. If these figures prove to be correct, the synodic period of this planet is about 19 years, and it may well be out of reach of our telescopes for long periods.

Some Events in 1978

ECLIPSES

In 1978 there will be four eclipses, two of the Sun and two of the Moon.

24 March – a total eclipse of the Moon, visible in Australia, Asia, Africa, and Europe.

7 April – a partial eclipse of the Sun, visible in South America and Southern Africa.

16 September – a total eclipse of the Moon, visible in Australia, Asia, Africa, and Europe.

2 October – a partial eclipse of the Sun, visible in northern Europe, Asia, and the Arctic regions.

THE PLANETS

Mercury may best be seen in northern latitudes as an evening star at greatest eastern elongation on 24 March, and as a morning star at western elongation on 4 September.

Venus is in superior conjunction on 22 January, and will be an evening star for the greater part of the year. It is at greatest eastern elongation on 29 August, and is at greatest brilliancy on 3 October. After inferior conjunction on 7 November it will be a morning star, reaching greatest brilliancy on 14 December.

Mars will be at opposition in Cancer on 22 January.

Jupiter is an evening star for the first half of the year, and after conjunction on 10 July will be a morning star. There is no opposition of Jupiter in 1978.

Saturn is at opposition in Leo on 16 February.

Uranus will be at opposition in Libra on 5 May.

Neptune remains in Ophiuchus and will be at opposition on 8 June.

Pluto will be at opposition on 5 April in Virgo.

Vesta will reach magnitude $+5\cdot5$ at opposition on 5 June, but is then low in the sky in Ophiuchus.

PART TWO

Article Section

The articles in this *Yearbook* are of more than usual interest, since several of them contain original research. Dr Martin Cohen describes his remarkable discovery of 'ice in space', while Dr van de Kamp, the distinguished Director of the Sproule Observatory in Pennsylvania, tells how it became possible to detect stellar companions which have yet to be actually seen – a field of research in which Dr van de Kamp in the leading pioneer. Dr Garry Hunt describes the recent findings with regard to that remarkable Jovian satellite, Io; Dr David Allen deals with interstellar dust, a subject in which he has made (and is making) major contributions. We are also delighted to welcome another article from H. G. Miles, one of our regular writers, concerning the Russian triumph in obtaining pictures from the surface of cloud-covered Venus. A new departure is the article by E. R. Turner, who has been carrying out research into the role of Flamsteed, the first Astronomer Royal, as Rector of Burstow in Surrey. Some of this work, too, is here published for the first time.

The Rev. John Flamsteed

ASTRONOMER ROYAL, AND RECTOR OF BURSTOW
E. R. TURNER

The name of John Flamsteed is one of the most honoured in the history of astronomy. He was the first Astronomer Royal; he was the author of a vitally important star catalogue, and he made many other notable contributions. The story of his achievements has been told many times. The present article deals mainly with Flamsteed in his other capacity – that of Rector of Burstow, near Horley in the county of Surrey.

Flamsteed was born at Denby, in Derbyshire, on 19 August 1646. His father, Stephen, was a maltster, and his mother, Mary, was the daughter of an ironmonger of Denby, John Spateman. John Flamsteed was sickly from birth, and when he was only two years old his mother died, leaving a month-old daughter. Two years later his father married again, but his second wife, Elizabeth *née* Bates, also died two months after giving birth to a daughter in 1654.

Flamsteed was educated at the Free School in Derby, and was obviously gifted, but he was constantly handicapped by ill-health; in 1660 he contracted a rheumatic disease, and was unable to attend school regularly for the next two years, though he continued his studies at home and concentrated upon Latin, mathematics, and astronomy.

Despite all these difficulties, he made rapid progress, and in 1669 he compiled an astronomical almanac which was accepted by the Royal Society. It was then that he first met Sir Jonas Moore, Surveyor-General of Ordnance at the Tower of London; he was presented with a micrometer, and the promise of lenses for a telescope. He met Isaac Newton at Cambridge, and in October 1671 entered his name at Jesus College. Three years later he took a degree of M.A. *per literas regius*, planning to take Holy Orders and settle at a small living near Derby, which was in the gift of a friend of his father's. Pending ordination he stayed in London as a guest of Sir Jonas Moore in his apartment at the Tower. He was ordained by Bishop Gunning in Ely House at Easter 1675.

Flamsteed came into real prominence by his connection with the 'longitude problem'. He pointed out the impracticability of the method submitted to King Charles II by a young Frenchman, and he stressed that the first essential was to compile a new star catalogue which would be more accurate than the best existing publication – due to the great Danish astronomer Tycho Brahe, who had lived before the invention of the telescope and whose measurements had therefore been carried out with the naked eye. The King appreciated the situation, and decreed that the stars must be 'anew observed, examined and corrected for the use of my seamen'. The Royal Observatory was set up in Greenwich Park, and Flamsteed was put in charge.

Unfortunately Flamsteed was expected to provide his own equipment, and his salary, after tax was a mere £90 per year. Thanks partly to Sir Jonas Moore he was able to assemble enough instruments to begin the work, but Moore died in 1679. Charles II himself died in 1685, but in that year Flamsteed was presented by Lord North to the living of Burstow, and his financial position was further improved when his father died four years later. When he was appointed to the living of Burstow, he succeeded Ralph Cooke who had been rector since 1637. Both Cooke and his wife, who died in the following year, are buried under the altar in Burstow Church, and it may have been at Mrs Cooke's funeral that Flamsteed first met their son, also named Ralph (of Lincoln's Inn). On 23 October 1692 Flamsteed married Ralph Cooke's daughter, Margaret, at St Lawrence Jewry in London.

Flamsteed's career at Greenwich was outstandingly successful, and resulted in the eventual publication of a vitally important star catalogue, though the work took a long time and was not com-

THE REV. JOHN FLAMSTEED

pleted until after Flamsteed's death in 1719. This story, too, has been told frequently, so let us return to Flamsteed's connection with Burstow.

The ancient Wealden parish of 4,850 acres extends from the Sussex-Surrey border to the south almost to Bletchingly in the north. It was served by the eleventh-century church until almost the end of the nineteenth century, when Outwood and Copthorne churches were built; it is recorded that the population in 1700 totalled 606 persons. The rural communities were mainly centred round the four moated manor houses; the oldest and most important of these was Burstow Court, adjacent to the church. The living at Burstow not only provided Flamsteed with an occasional respite from his arduous duties at Greenwich, but also a very welcome addition to his income. Old accounts indicate that the direct financial returns were very small after paying all expenses, including the income for his curates. However, in addition to the tithes, he did receive goods in kind, all of which materially assisted in financing his work at the Observatory.

References to the Burstow living are found in some surviving original letters which passed between Flamsteed and his curates. These are held in the archives of the Royal Greenwich Observatory at Herstmonceux Castle. (By kind permission of Dr Alan Hunter, Director of the Observatory until the end of 1975, copies of these have been made available to me by Mr P. S. Laurie, without whose help and encouragement this history would have been very incomplete and unauthenticated.)

From these letters, and entries in the church registers, it appears that Flamsteed usually visited his parish twice a year. In letters to Isaac Newton, dated 28 March 1700, and again in July 1705, he says, 'I am going into Surrey for six or seven days', and 'I must go into Surrey to reap my harvest, as I usually do at about this time'. It is revealed that he took a great interest in the details of the conduct of his parish affairs and the care and maintenance of the 100 acres of Glebe.

Communications were undoubtedly difficult. There is a reference in 1700 to the fact that 'the penny post comes something quicker than the General Post'; and writing to his curate Timothy Stileman, he says: 'Last Monday I received two letters from you. The *idle post girl* had kept one of yours by her for some days to save herself the labour of a journey to the Observatory with it alone and brought it, your second and others together, two days after your Latin letter arrived.' Letters for Burstow were dealt with by the Postmaster at Reigate, and arrangements had to be made for them to be sent on to Horley for collection.

The network of roads we know today did not exist. The very few tracks and bridleways were mainly for local use, and were very hazardous throughout the Weald in winter. It was not for another hundred years that highways to take wheeled vehicles were put in hand to link towns and villages. Unfortunately, no details have come to light of the route and conditions experienced by Flamsteed on his many journeys between Greenwich and Burstow. Occasional references are make in his letters, but hacking for more than thirty miles under the prevailing conditions must have been very arduous for a bulky man in poor health. At the age of sixty he wrote to Mr Stileman and rebuked him for his absence from his duties at Burstow without authority or notification, when he went to Oxford to take his degree. He goes on to say: 'I write to acquaint you of the badness of the ways. I dare not yet take my journey, for I am scarce yet well of my illness and the pains of my foot. I am much afraid that the cold and the wet which I must necessarily meet below the hills may cause a relapse worse than the first distemper. I shall therefore save my journey for a fortnight or three weeks after the holiday.' Also: 'Advise Michael Todd, Mrs. Smith and my tenants I expect my rent when I come down.'

The reference in a later letter to goods sent by carrier may indicate that the road from Lewes, via East Grinstead and Godstone, was in limited use. On 22 November 1706 he writes: 'Yours of the 9th inst. I received as I was going to London in the coach on Wednesday was seven-night with my wife. We sent the boy back that night to look after the fruit, but not returning ourselves till Fryday night, I could not give you an account of its arrival before this. We found it safe and on opening it, the apples on the top were very good but the pears underneath were injured by reason of the weight lay upon them or that they had been too long put up before they were brought by the carrier.'

Flamsteed also reprimanded Stileman for not visiting him at the Observatory on his journey back from Oxford. He says: 'Mr. Green was lately in town – if the ways are not so abominable bad but he can travel them, I suppose it will not be worse for you than to him and you are not so heavy as he is, this is no good excuse.' It was part of the curate's duties to care for the Glebe lands, for Flamsteed goes on to say: 'I am very sorry that the thatching of the barn was so long delayed. I fear the straw was injured by too much wet. This ought to have been done before you went to Oxford . . . My respects to your sister, uncle and aunt . . We are told King Augustus of Poland marches for Italy with 1900 men to assist Prince Eugene.'

Goodman Chart, the Parish Clerk, was also involved, for in a duplicate letter to the curate he was told: 'I think it advisable to let my copse stand for another year before I cut it and therefore desire you to make up the gaps round it very well to prevent John Steers cattle from getting into it and making any further waste.' In a postscript, he continues: 'I think it my turn to choose the Churchwarden, pray look into your book and see if it be, let me know by your Reigate post and who of my neighbours who have been longest out and I will order you who to pitch upon.'

Two more letters in that year introduce a note of sadness, and the problem of keeping a curate to look after his benefice. From the Observatory he wrote to his friend, Abraham Sharp, of Little Horton, Yorkshire: 'My friends here die apace. I grow gouty and the pains of my feet hinder me from stirring much abroad so I am confined in a manner to my business.' And in August he wrote: 'Mr. Witty I have dismissed from the Observatory and he is now chaplain and companion to a young gentleman in Hampshire on better terms than I could afford him.' The problem of suitable replacements also occurred at Burstow, and is referred to in a letter from Bishop W. H. Kennett, who signs himself as 'your affectionate friend'. It runs: 'November 1st 1715; I have been so anxious to provide a safe and honest curate for you and I think at last I have fixed upon a man against whom I know of no objection. I sent to Peterborough for him where I knew him well to be a sober quiet man, and now happens to be lately out of business; I have proposed your curacy to him at the terms of forty pounds per annum and perquisite of a small parish. He is one Mr. Francis Peck, a single man of our side with good modest sense and very sufficient learning. If you let me know when he shall attend you and go down to residence, I will send for him and I believe he will fully answer your desires. I am now in waiting at court and should be glad you would dine with me.' There follows a letter of introduction, with the comment: 'There is but one objection, he is only Deacon but that defect shall be supplied by my care of getting him into Priests orders as soon as maybe.'

It would appear that the Rector was not always received with open arms when he visited Burstow. The year following his installation, Mr. Peck writes: 'I had the favour of your kind letter, and the next day brewed the drink you ordered and gave the Clerk directions to acquaint everybody of your intended resolution to be at Burstow on this day sennight, and what you expected from them, but I find they murmur and grumble at ye proposal. When you come down pray please let 'em put up the Fryingpan and Warmingpan you promised me.' . . . 'Mr. Clark began to teach school at his house the Monday after Easter. Mr. Rebow allows him for two, and I for one poor child to learn to read, and he hopes you will send him some more.' The Rebow family were substantial landowners in the parish; they had come to Smallfield in 1703.

In the same year, 1716, Flamsteed wrote two letters to his friend, Mr. Sharp, which fix the date for the alterations to the vicarage. In August he wrote: 'I have now completed the 70th year of my age; a little business now lies heavily upon me than a great deal did formerly. My Parsonage House is part of it, plucked down and rebuilding. . . My strength impairs daily, so that I can now walk but once a day to church on Sundays. My memory

THE REV. JOHN FLAMSTEED

and reason continue still – praise to God.' In the second letter, 28 December, he wrote; 'I have rebuilt three quarters of my Parsonage House at Burstow at about £120 charge, so that it is now the best in the country. I am feeble and have swimming in the head. I praise God for the health I enjoy.'

Despite all his suffering, he invariably ended his correspondence with thanks to God for his ability to carry on. In what was probably his last letter to his friend, dated 13 September 1718, he wrote: 'I have voided another small stone yesterday, and without much pain. I am now complete 72 years of age, and entered on my 73rd. I thank God I have my health well for my years, and I doubt not he will continue it that I may finish what I have in my hands.'

The name of Flamsteed is perpetuated in the memorial east window of Burstow Church, and in the sculptured plaque presented by the Turner family to mark the tercentenary of his appointment as Astronomer Royal. In his will, dated 1717, Flamsteed left the income from £25 for two coats for two old poor folks; and his wife Margaret, in her will dated 1728, left the income from £25 for two gowns and petticoats for two poor widows. These gifts have now been amalgamated with other bequests dating from 1626, and are today managed by the Trustees of Burstow Parochial Charities.

John Flamsteed was a great astronomer. We should never forget that he was also a conscientious Rector of Burstow.

The Surface of Venus Revealed

H. G. MILES

To the non-astronomer it must seem very strange that man is able to glean a tremendous amount of knowledge of the outermost regions of the universe, yet be so blatantly ignorant of many of the basic facts about our nearest neighbours in space. Many of the so-called facts about these nearby objects, collected over several centuries, have been shown to be wrong or highly misleading if one accepts the information gained from a few probes carrying relatively simple experiments. This information, together with studies in the new astronomies using infra-red, ultra-violet, and radio wavelengths, have replaced much of the pre-space age data, but they themselves have produced a picture which is by no means complete. I suspect that much of the present-day accepted facts will, within a few years, be shown to be inaccurate.

In the case of the exploration of the Moon and Mars, there has been a gradual development of our knowledge, and the process can be likened to observing them with an ever-increasing magnification, with each step producing new information but at the same time fitting it into the generally expanding picture. With Venus, however, because of our virtually total inability to penetrate the cloud layer from Earth-based studies, every discovery from a probe becomes a sudden jump forward. It is true that Earth-based radar studies have penetrated the cloud cover and produced maps of the surface, and that ultra-violet studies have produced information on the upper regions of the atmosphere. It is also true that the conditions existing on Venus were predicted before the sending of spacecraft; but it must also be recalled that there were many other theories and ideas which appeared, at that time, to be

THE SURFACE OF VENUS REVEALED

just as reasonable. The arrival of each probe has tended to become the clearing-house for theories, and we are left with the impression of knowledge arriving in discrete packages.

Since 1962 Venus has been visited by spacecraft from both the Soviet Union and the United States. The last American probe, *Mariner 10*, provided much information on the upper atmosphere (see *Yearbook of Astronomy* 1975). The Russians, on the other hand, have concentrated on the problems associated with the lower atmosphere and the surface. The latest probe reached the surface, took panoramic photographs of the landscape, and then transmitted the details back to the Earth. To appreciate fully the tremendous technological feat achieved, it is necessary to review the conditions existing on the planet in the light of present-day knowledge.

At one time Venus was thought to be similar to the Earth, because it is only about 5 per cent smaller; yet, because it is about 28 per cent nearer to the Sun, any further similarities would be hard to find. It rotates on its axis in 243 days in a sense opposite to its orbital motion round the Sun and to all the other major bodies in the Solar System. It has a negligible magnetic field, and hence does not possess the equivalent of the Earth's Van Allen radiation belts. It is covered by opaque clouds, which have prevented astronomers from seeing any of the lower regions. The upper layers of this cloud cover, when observed in ultra-violet radiation, show features which rotate in four days. For comparison, some upper regions of the Earth's atmosphere rotate faster than the surface, but by only a very small percentage – nothing like the factor of 60 occurring on Venus. The surface pressure on Venus is in the region of 100 bars, i.e. 100 times greater than that existing on the Earth. It consists of about 97 per cent carbon dioxide with only small amounts of nitrogen and oxygen. This raises the interesting question as to where all the carbon dioxide has come from. On the Earth, if we consider all the carbon dioxide locked up as carbonates in the sedimentary rocks in addition to the free gas, there is reasonable agreement in the total amount. The surface temperature on Venus of over 400°C is far too high for many complex molecules to exist. It seems that this high temperature exists during the night-time as well as the day. To crown the idea of a totally inhospitable world is the information obtained from the upper region of the atmosphere. These layers consist of droplets of sulphuric acid, and other corrosive acids are suspected at other levels.

It has been into this hostile environment that the Soviet Union has sent several probes. At the time of the first of these probes, the atmospheric conditions were unknown, and much of our knowledge of these regions has been due to a systematic attack on the problem. The first controlled entry into the atmosphere (as opposed to a direct crash collision course) took place in 1967 when *Venus 4* descended by parachute for 90 minutes, giving details of temperature, pressure, density, etc. It suddenly stopped transmitting, and it is now thought that this occurred when the capsule collapsed due to the high external pressure exerted by the atmosphere. In 1969, two further probes, *Venus 5* and *Venus 6*, suitably strengthened, descended by parachute, sending back information for just under an hour. On these occasions, gas analysers were used to determine the composition of the atmosphere.

Venus 7, launched in 1970, extended the study of the atmosphere and became the first probe to transmit from the surface of the planet. The distinction of being the first probe to send details of its surface fell to the next probe, *Venus 8*, in 1972. Previous probes had studied the night side only, and it was therefore a natural development of a systematic study of the plamet that a probe should descend on the sunlit side, to see if conditions were the same. *Venus 8* drifted downwards for just under an hour, and showed that conditions were roughly the same. Wind speeds were also measured. Analysis of the character of the radio signals sent back to the Earth as the capsule descended through the atmosphere, the signals being reflected from the planet's surface, suggested that the surface consisted of loose rocks of density slightly less than 1·5 g per cubic centimetre. Analysis of the rocks indicated that the area on which the probe landed was similar to terrestrial granite.

THE SURFACE OF VENUS REVEALED

It was against this background knowledge that *Venus 9* and *Venus 10* were launched to continue the study of the planet. The first, *Venus 9*, was launched on 8 June 1975, and six days later its twin was sent on the four-month journey to the planet. From statements issued by the Soviet Union, it was clear that the primary objective of these probes was to reach the surface. A rapid descent through the hostile atmosphere was essential, and so the parachutes were redesigned for this purpose. It may be recalled that with *Venus 4* the descent was very slow; consequently the area of the parachutes for *Venus 5* and *Venus 6* was reduced from 50 to 15 square metres. Even this was not sufficient for a rapid descent, and so for the subsequent probes the area was reduced even more to 2·5 square metres. With this small area the probes reached the surface within the hour.

On 22 October, it was announced that 15 minutes after soft landing on the surface, *Venus 9* had sent back the first picture of the surface of the planet. The craft had landed in a region covered with boulders of various sizes. Dr Mikhail Marov, head of the team responsible for the experiment, said that he thought there may have been difficulties in obtaining clear pictures because the solar light might have been diffused to a large degree by the dense atmosphere. This turned out not to be the case, and the clarity of the pictures enabled the investigators to indicate that the stones were relatively young on account of their angular nature. The capsule was in fact fitted with floodlights because of the possibility of the surface being dark, so that it would have been possible only to spotlight isolated areas.

The orbital techniques employed with *Venus 9* were different from those used on the earlier probes. On arrival in the vicinity of the planet the spacecraft was put into orbit round Venus, with a minimum height of 1,300 km and a maximum of 112,000 km, the period being 48 hr 18 min. The descent capsule had in fact been separated from the spacecraft two days earlier, and then continued independently on a course which took it directly into the atmosphere. Aerodynamic braking reduced the speed from 10·7 km sec^{-1} to 250 m sec^{-1}, and then the parachute system was

Descent module identical to that of Venus 9

brought into operation. As it descended, the cloud layers were investigated. When it reached a height of 50 km above the surface, the parachute system was detached and the probe continued to descend by free fall. Attached to the top of the capsule was a round metal disk which acted as an air brake. This, together with the dense atmosphere, allowed the probe to descend at a reasonable speed, but at the same time not too high a value so that the capsule would be damaged on impact with the surface. Another structure underneath the capsule helped to cushion the force of the landing, which took place at a speed of between 7 and 8 metres per second. During the time of descent and for 53 minutes after landing, the data collected were sent to the orbiting spacecraft and then re-transmitted back to Earth. At the *Venus 9* landing site, estimated to be 2500 metres above the Venusian 'sea-level', the

THE SURFACE OF VENUS REVEALED

surface temperature was 485°C and the pressure 90 Earth atmospheres. During its active life on the surface, in addition to the sending of the photographs, information was relayed back on the nature of the surrounding rocks.

Venus 10 behaved in a similar manner on arrival in the vicinity of Venus, and sent a capsule to the surface of the planet at a point about 2300 km from the landing site of its twin. The spacecraft itself went into an orbit varying between 1400 and 114,000 km from the surface, with an orbital period of 49 hr 23 min. The capsule descended through the atmosphere in 75 minutes, and transmitted information from the surface for a further 65 minutes.

Surrounding area of Venus 10 landing capsule

At the landing site the pressure was 92 Earth atmospheres and the temperature 465°C. The wind-speed was recorded as 3·5 metres per second (about 7 miles per hour).

Venus 10 also transmitted a picture of the surrounding area. The terrain appears to be entirely different from the *Venus 9* landscape. In contrast to the 'young' appearance of the earlier photograph, this area showed a landscape typical of old mountain formations. The rocks were rounded, and resembled huge pancakes with sections of cooled lava or debris of weathered rocks in between. This initial assessment was given by Boris Nepoklonov of the Institute of Geodesy, Aerial Survey and Topography.

Soviet geologists have indicated that it is difficult to explain the rock-strewn surface without thinking that the boulders were the result of debris from volcanic explosions of from some other explosion process. The boulders resemble closely the fractures of sedimentary rocks such as shales. The sharp edges of the rocks seen on the *Venus 9* photographs suggest an active planet and could possibly be taken as an indication of tectonic activity.

During their travels round the planet, the orbiting spacecraft have been sending back photographs of the clouds taken in ultraviolet light. The results are supplementary to the excellent ones obtained from the *Mariner 10* fly-by in 1974. The initial photographs covered strips about 1500 km wide, and showed evidence for the equatorial current and spiral structure towards the poles. Unfortunately these photographs are limited to the upper layers of the extensive cloud system existing in the atmosphere of the planet. The big questions are 'How thick is the cloud layer?' and 'What do the clouds consist of?' The descending capsules have provided some of the basic information, but there are still many interesting facts to be ascertained. The clouds extend over a height range of 30 to 40 km, not as a single thick belt but as a series of layers. The lower layers are about 30 to 35 km above the surface of the planet. The uppermost layers seem to be composed of droplets of concentrated sulphuric acid, with some hydrochloric and hydrofluoric acids also present. Because of the high surface temperature, many compounds would tend to vaporize, resulting in gases such as bromine and iodine migrating into the atmosphere. These will be in relatively small amounts compared with the small percentage of nitrogen and oxygen and main constituent, carbon dioxide.

A surprising fact deduced from the information sent back is that in spite of the relatively high landing speed, no dust-cloud was sent up from the surface. This suggests that Venus has a dust-free surface. The atmosphere appears to be completely free of dust absorption. This can be implied by the fact that in the photographs, the horizon can be seen as a very sharp line.

Once again, with a few facts under their belts, the experimenters are busy with theories trying to explain these observed facts. Unjustified extrapolation and interpolation permit many reasonable ideas which could give a completely erroneous picture of the

THE SURFACE OF VENUS REVEALED

conditions existing on the planet. Consequently it can be forecast with a fair degree of confidence that many quite reasonably sounding theories will be completely thrown to the winds when the next probes visit the planet and send back some more facts. Obviously the Soviet Union has not finished its study of Venus, and will send further probes during the next launching window. We must not forget, however, that the Americans have extensive plans for further studies of the planet. In 1978, they hope to launch two spacecraft, one to orbit Venus and study the planet at close quarters and the other to carry four probes which will enter the atmosphere. Whoever sends the probes, one thing will be assured. There are a lot of surprises in store.

Io, the Anomaly of the Solar System

GARRY E. HUNT

1. *Introduction*

Jupiter, the largest planet in the Solar System, is surrounded by an extensive system of 13 satellites. The innermost group number five. The closest to the planet is the small moon Amalthea, which is about 150 km in diameter, and then come the four brightest satellites of Jupiter discovered in January 1610 by Galileo using one of the first telescopes ever constructed. In his book *Sidereus Nuncius* (The Starry Messenger), he reported his amazing discovery by saying '. . . But that which will excite the greatest astonishment by far and which indeed especially moved me to call the attention of all astronomers and philosophers, is this, namely, that I have discovered four planets, neither known nor observed by anyone of the astronomers before my time. . . .' With a mind towards a future position in the Tuscan Court, Galileo named them 'Sidera Medica', but his rival, Simon Marius gave them the names from mythology by which they are known today, Io, Europa, Ganymede, and Callisto. In Table 1 we have compared some of the physical properties of these satellites with those of our Moon.

TABLE 1
The Inner Satellites of Jupiter compared with the Moon

	Radius (km)	Density ($g.cm^{-3}$)	Distance from primary ($10^3 km$)	Sidereal period (hr)	Geometric albedo	Opposition magnitude
Amalthea	75	?	181	12	?	13
Io	~1810	3·52	422	42	0·6	+4·8
Europa	~1480	3·3	671	85	0·7	+5·1
Ganymede	~2600	1·95	1070	172	0·4	+4·4
Callisto	2450	1·6	1883	401	0·2	+5·4
Moon	1780	3·34	384	656	0·1	(−12·7)

IO, THE ANOMALY OF THE SOLAR SYSTEM

We see that Io is comparable with the Moon and Mars, while Ganymede is one of the largest satellites in the Solar System. It is interesting to note that the densities of these satellites decrease in magnitude as we move away from the parent body, Jupiter, just as the planets in our solar system as we move outward from the Sun. Jupiter and the Galilean satellites represent a mini-solar system, and understanding their formation may therefore hold important clues to the formation of the Solar System.

The satellites are quite bright, and could be seen with the unaided eye on a clear night were it not for the blinding glare of Jupiter, which is about a thousand times brighter. However, the use of a pair of binoculars soon resolves the difficulty.

The satellites are eclipsed when they pass into Jupiter's shadow, and both they and their shadows are seen passing across the planet's disk when they come in front of it. On almost any night one can observe one or more eclipses or transits of the large satellites. The timing of the eclipses is the most accurate means of studying the motion of the satellites in their orbits and determining their perturbations.

The timing of the eclipses of Jupiter's satellites led to the first detection and actual measurement of the velocity of light by Ole Rømer in 1675. He noticed that when eclipses of Jupiter are viewed across the Earth's orbit they seem to occur relatively later than when Jupiter and the Earth are on the same side of the Sun. The difference represents the time taken by light to cross the orbit of the Earth.

The Galilean satellites, then, have long been quite simple to observe and study, and it might be thought that our understanding of them should have been congealing in an orderly fashion. But recent observations of these bodies show that they exhibit puzzling properties, while Io in particular displays some of the most bizarre phenomena found in the Solar System.

The five inner Jovian moons reside within the planet's magnetosphere, which is a region of trapped radiation around the planet. The orbits of the rest of the satellites are generally beyond the magnetosphere, just as our Moon revolves beyond the equivalent region around the Earth. Each of the inner satellites can intercept charged particles and thereby remove them from the population of the radiation belts. As a moon revolves in its orbit it can sweep a corridor clear, while itself acquiring intense radioactivity. During the recent Pioneer 10/11 mission it was found that both Io and Europa have a significant interaction with the radiation belts. This particle sweeping effect of these inner satellites is thought to have been of fundamental importance for the survival of spacecraft and, therefore, for the success of the Pioneer missions. Another strange property of Io was found by Earth-based radio astronomers, who observed a strong correlation between the position of Io in its orbit and bursts of radio noise from Jupiter, which are equivalent in energy to those of thermonuclear devices. Apparently the radio noise is generated when Io disturbs the magnetic field of Jupiter in its swing around the planet.

Visual astronomers had discovered a further puzzling feature. Some reported an increase in brightness of 0·1 magnitude lasting as long as 15 minutes upon emergence of the satellite from Jupiter's shadow, while other observers failed to detect any change in brightness. Does this suggest that the satellite may have an atmosphere? Certainly one would not expect a planetary body with such a low surface gravity to possess a substantial one, so that the discussions of an atmosphere on Io have always been controversial.

2. *Atmosphere and Ionosphere*

But Io does have an atmosphere, and an ionosphere too. It is, therefore, the smallest body in the Solar System with such features. The ionosphere was first detected during the Pioneer 10 fly-by of Jupiter in December 1973, and was found to extend to some 700 km above the surface, with a peak electron density of about 6×10^4 electrons cm^{-3} at an altitude of between 60 and 140 km on the day-side. A thinner and less dense region was observed on the night-side with a peak density of about 9×10^3 electrons cm^{-3} at an altitude of 50 km. The diminished night-side ionosphere indicates that this region is produced by the action of solar radiation on the day-side, and then decays during the 21-hour Io night. The large

IO, THE ANOMALY OF THE SOLAR SYSTEM

diurnal variation implies a short ion lifetime, while the large electron density implies a long lifetime or an unusually large production rate. This matter had not been resolved yet. It is possible to explain the long lifetime by assuming an atmosphere of pure neon, with the dominant loss at the surface. But such an atmosphere is hard to accept. Moreover, there is no evidence for abundant neon in any other atmosphere.

The Io atmosphere, with a surface pressure of between 10^{-8} and 10^{-10} bars, is composed of exotic constituents; sodium, potassium, hydrogen, and possibly ammonia and nitrogen. The measurement of sodium emission from the satellite's atmosphere was the first positive evidence. Subsequent studies showed that the emission also came from an extended space around Io, with fainter emissions originating in remote parts of Io's orbit and beyond. These studies suggest that the sodium in Io's vicinity may be partitioned into three regions shown in Figure 1. Region A is spatially coinci-

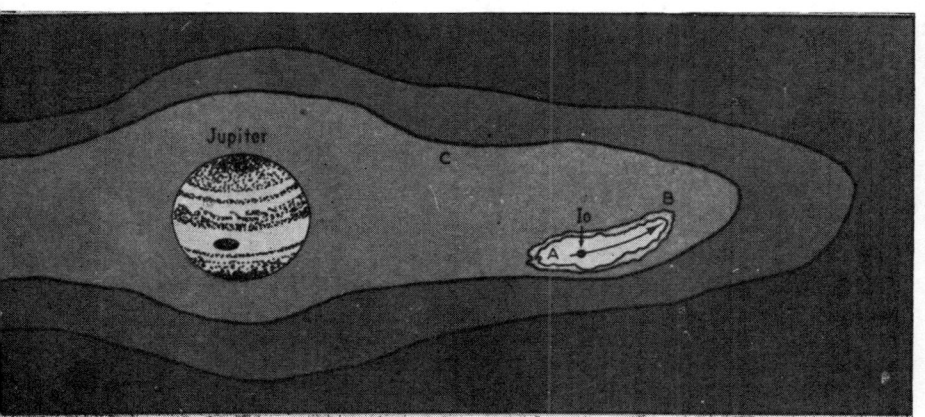

Figure 1. The geometry of Io's cloud

dent with Io's visible disk, and corresponds to sodium in Io's bound atmosphere. Region B has direct association with Io and its immediate vicinity. Region C covers the outer region and extends even over the poles of Jupiter. The column abundance of sodium is estimated to be roughly 10^{11} cm^{-2} in this extended cloud (region B) and 10^{13} cm^{-2} in the atmosphere of Io (region A).

In addition to the sodium cloud, there is around Jupiter in the orbital plane of Io an extensive cloud of atomic hydrogen, detected by the UV photometer on Pioneer 10. The mean diameter of the torus is about equal to the diameter of the orbit of Io. The torus is not complete, however, but seems to extend for about 60° on either side of the satellite. The lack of emission from that portion of the cloud which happened to lie in Jupiter's shadow suggests that the cloud diameter is at most equal to that of Jupiter, and that resonance scattering of sunlight is the primary excitation mechanism.

A torus of this type is thought to be formed by atoms which can escape from the satellite but do not possess sufficient energy to escape from the vicinity of the planet owing to its large gravitational field. Consequently the atoms are bound in closed orbits until lost by ionization or recapture, and tend to produce a toroidal-shaped cloud whose density is determined by the ionization losses. A similar mechanism is thought to be responsible for the proposed torus around the Saturnian satellite, Titan, which we discussed last year. The cloud around Jupiter has a brightness of about 10^4 Rayleighs, composed of roughly 10^{33} hydrogen atoms with a mean lifetime of $\sim 10^5$ seconds. This is slightly longer than Io's orbital period.

3. Surface Properties

But is Io the source of the toroidal hydrogen? The escape of hydrogen from Io poses no problem, since the Jeans escape from an exosphere as cold as 200°K is still very efficient. In order to maintain the cloud, the satellite must be supplying hydrogen at a rate of 10^{11} atoms cm^{-2} s^{-1}. An escape flux of this magnitude would exhaust the entire atmosphere of Io in a few years. Is the surface of Io abundant in hydrogen-rich material?

There are several other puzzling questions regarding the surface of Io that must also be taken in account. Firstly, why does Io have such strong sodium emissions? Also, why is Io as bright as if it

IO, THE ANOMALY OF THE SOLAR SYSTEM

were covered by ice, yet shows no ice absorption features in its spectrum? And why has Io dark poles? Observations by Barnard at the turn of the century suggested Io had reddish caps have been confirmed by more recent studies, which suggest that the equatorial region may be yellow in tone.

Any explanation for Io's surface composition must explain the spectral curves of Io and the other Galilean satellites in a manner which is consistent with their cosmochemical setting in the Solar System. The multi-colour photometry and moderately high-resolution spectroscopy of the satellites have given a relatively clear description of the reflection properties of the surfaces of these bodies between 0·3 and 4 μm. In general, the satellites all show a very low UV albedo, although this effect is less pronounced for Callisto because its albedo at longer wavelengths is also low. This feature is typical of pure frosts, leading to the suggestion that the probable surface frosts contain admixtures of other compounds. The change in shape of the albedo curves from Ganymede to Io may be related to the interaction of these satellites with the Jovian magnetosphere. This has been suggested as a possible source of temperature-sensitive free radicals in the surface of Io, and among the suggested compounds have been NH, CH, CH_3I, $(NH_2 NH)_n$, CH_3, CS, polymers of HCN, C_2N_2 and $HCN.NH_3$; others suggested are NH_4SH, radiation-damaged ammonium hydrosulphates, and ultra-violet–irradiated ices, but many of these compounds can be eliminated, since they do not satisfy the reflection spectra over the complete UV – IR range.

It is thought that the solid material in the preplanetary nebula at Jupiter's heliocentric distance was composed of a mixture of carbonaceous chondritic material and ices. The differences in density between the Jovian satellites may then be due to an initially high surface luminosity for Jupiter. This would prevent the incorporation of large amounts of H_2O in Io, which has the highest density (Table 1.). Europa, by comparison, has a lower density and strong H_2O ice bands in its reflectance spectrum. Ganymede clearly contains a very large portion of H_2O ice as suggested by its low density of $\sim$ 1·94 g cm^{-3}. Io may never have had large amounts of ice, yet apparently was not totally devoid of water like our Moon. As a result of internal degassing, much of Io's water may have seeped to the surface, where it would quickly evaporate as a result of the satellite's proximity to Jupiter and its low surface gravity. The high density of Io and absence of H_2O ice bands in the surface spectrum may mean that this process is essentially completed. Perhaps, then, the internal processes are considerably less advanced on Europa and Ganymede, since these satellites are apparently very different from Io.

The surface of Io may therefore involve evaporite salt deposits, rich in sodium, sulphur, potassium, derived from the leaching of carbonaceous chondritic material. This would explain the brightness of the planet and provide a possible source of the sodium.

4. *Atmospheric Evolution, Stability, and Escape*

But how is the atmosphere created? Meteoritic material has been suggested as a source of Io's sodium, and this mechanism is thought to be responsible for the sodium in the Earth's atmosphere.

Measurements of micrometeoroids in the Jovian environment by the Pioneer 10/11 spacecraft indicate that the mass flux of meteoroids intercepted by Io could be two orders of magnitude higher than that for the Earth ($\sim 2 \times 10^{-16}$ g cm^{-2} s^{-1}). For a sodium concentration of 0·66 per cent by weight, the net sodium source is approximately 3×10^6 atoms cm^{-2} s^{-1}. The sodium atoms can be released from the meteorites by evaporation at the moment of impact, but this mechanism does not explain why sodium has not been observed on any of the other Galilean satellites. Alternatively, the sodium could be created by sputtering after the material has settled on the surface. This then requires the sodium content in Jovian meteorites to be higher by at least a factor of 10 than that typical of chondritic meteorites.

It is difficult to release sodium from the surface, since both metallic sodium and most common sodium compounds have low vapour pressures at Io's surface temperature of 140°K. But Io resides in the intense Jovian magnetosphere. Consequently, the observed sodium in the atmosphere is probably sputtered from Io's surface

IO, THE ANOMALY OF THE SOLAR SYSTEM

by the charged particle bombardment, and these atoms populate the cloud observed beyond the satellite. This mechanism may also beresponsible for the potassium emission observed. The previously unexplained coloration of the surface must therefore result from the Jovian magnetospheric fluxes being strongest at the poles of Io. But we should not exclude Ionian surface activity as a further contributing mechanism for the sodium cloud. Although sputtering is the more likely method, it is difficult to completely understand how it could have continued for the life of the Solar System.

The presence of ammonia on the surface and in the atmosphere of Io is still controversial, since there are no ammonia absorption features in the satellite's infra-red spectra. However, an important reason for assuming an ammonia atmosphere is the need to supply a huge escape rate of hydrogen. If Io's atmosphere contains $\sim 10^{18}$ molecules cm^{-2} of ammonia, its photolysis would lead to the production and escape of hydrogen at the rate of 2×10^{11} atoms cm^{-2} s^{-1}. This process represents a net absorption of solar energy and heats the upper atmosphere to around 500°K. Nitrogen would be an important minor product, which when excited by an auroral mechanism could interact with sodium to cause sodium emission. The sodium photo-ionized by solar radiation may be the dominant source of ionization. The auroral, possibly resulting in the sodium emission, may draw energy from the current arches to Jupiter, so that significant localized atmospheric heating may result providing for thermal escape of exospheric sodium into the surrounding cloud.

The Io-controlled radioactivity of Jupiter is highly sporadic, so that a synoptic patrol of the brightness of the sodium emission may be used to examine the various roles suggested for the high-energy particles. Such a study has found that resonant scattering of sunlight is primarily responsible for the brightness of the sodium cloud. Auroral processes may still be important for the near surface emission and for heating sodium to its high kinetic temperature. The stability of the cloud over a 43-night observing period implies a remarkably steady source of sodium, since the mean lifetimes of steady sodium is less than 16 days.

5. *Conclusion*

Io is an anomalous fragment of primordial material bathed in a violent electrical environment which modifies and erodes it. The Jupiter–Io system represents a unique environment where two atmospheres are mutually interacting on a large scale through a magnetic field. We must note the truly unique nature of sodium on Io, since it is the only Galilean satellite to possess this emission. Only potassium and sodium have been detected on Io so far. The absence of other metallic emissions and the absence of sodium emission from the other Galilean satellites could then be used to provide useful constraints on the chemical nature of their surfaces or the mechanism for releasing the metallic atoms from their bound state.

There is no conclusive observations yet of atmospheres on the other inner Jovian satellites. Measurements from the Pioneer 10 UV photometer suggest that there may be a hydrogen toroid in the orbit of Amalthea (JV), in addition to the Io toroid. If it is real, the toroid may be due to charged particles recombining at the surface of Amalthea, in a similar way which causes the Ionian hydrogen. Although Europa also plays an important role in sweeping up charged particles in a manner similar to Io, there is no evidence yet of any atmosphere.

Correlated ground-based observations, as well as *in situ* measurements from spacecraft, are still required if we are to understand this strange body Io, and its interaction with the Jovian environment. In 1979 the two Mariner spacecraft will fly through the Jovian magnetosphere at distances of $\sim 6R_J$ from the planet. This will enable observations to be made of the Ionian-Io atmosphere, ionosphere, and torus and high-resolution studies of the surfaces of the satellites. These studies will be critical for our quest to understand the evolution of satellite atmospheres, and may help to explain why Io is the anomaly of the Solar System.

The Unseen Stellar Neighbourhood

PETER van de KAMP

1. *Introduction*

In 1938, I published an article entitled 'The Invisible Universe' (van de Kamp 1938). It dealt with two aspects of the optically unseen, namely (1) the interstellar material which in low galactic latitudes severely limits our vision, leading to dim-out and blackout effects; (2) unseen companions of stars whose existence is revealed and studied by the gravitational influence of these companions on the visible primary component of that binary system.

A great deal has happened since 1938. Up to the advent of radio astronomy the unseen galactic material was revealed through the selective absorption and the scattering of light, and the resulting apparent distortion of galactic structure (*loc. cit.*) Radio observations are now penetrating the obscuring material, and have aided in outlining both the disk and the spiral structure of our Galaxy primarily through a study of neutral hydrogen. The invisible galactic universe is gradually yielding its secrets up to distances of tens of thousands of light years, not to mention the radiostudies of distant galaxies and quasars, millions of light years away.

As to the study of unseen companions of stars, there has been progress, too, but it has been limited to our immediate galactic neighbourhood, well within 100 light years. The high percentage of binary (and multiple) stars in our neighbourhood strongly suggests, and it has been confirmed observationally, that many stars, known as single, have as yet unseen companions. Also unseen companions are found in known double and multiple systems. Truly single stars appear to be a small cosmic minority.

The discovery of comparatively massive unseen companions by

the spectroscopic method has been carried out for a long time. This approach generally leads to the discovery of short-period binaries with high orbital velocities. Astrometric methods are particularly suited for the discovery of unseen companions with long periods and correspondingly large amplitudes (van de Kamp 1975b).

While a single star moves in a linear path with uniform velocity, the components of a binary system describe elliptical paths ('Kepler motion') around their centre of mass. The latter moves as a single star, with a uniform rectilinear motion. As a result each component describes a wavy motion which is the resultant of the motion of the centre of mass and of the orbital motion of the component.

The seen stars in our immediate neighbourhood, say within 10 parsecs, fall for the most part on the main sequence while several are white dwarfs. The overwhelming majority of these main sequence stars are faint red-dwarf stars. In our neighbourhood, there must be many, at first sight single stars, which in fact have a secondary component, not yet seen either because of too close proximity to the primary, or because it is intrinsically very faint or emits no light; both factors may play a role. However, the unseen companion may have sufficient mass to reveal itself astrometrically through its gravitational interaction with the visible primary, which as a result appears to describe a wavy motion. The amplitude and period of the so-called *perturbation* yield information about the mass of the unseen object.

The search for and discovery of unseen astrometric companions is the subject of the present article. Some individual results as well as tentative statistical conclusions will be presented.

2. *Historical*

The methods and techniques of discovering unseen companions of stars have a long history. Perturbations in 'single' stars were discovered as early as the beginning of the nineteenth century. In 1844 Bessel announced the 'variable proper motion' in the paths of Sirius and Procyon, both nearby stars at distances of 8·6 and 11·0

THE UNSEEN STELLAR NEIGHBOURHOOD

light–years respectively. Bessel based his conclusion on transit-circle observations. The companions were seen with large telescopes in 1862 and 1898 respectively. The orbits of the Sirius and Procyon systems have periods of 50 and 39 years respectively; the faint companions are white dwarfs.

Perturbations in the relative Keplerian motion of known double stars, pointing to a third unseen object, were found for Xi Ursæ Majoris (Nörlund 1905) and Zeta Aquarii (Strand 1942). These companions have not yet been seen, but most likely they are dwarf M stars. One of the most challenging perturbations (Seeliger 1888) is that of Zeta Cancri C, the distant companion of the well-known binary Zeta Cancri AB. The unseen companion, Zeta Cancri D, proves to have a mass close to that of the Sun, but thus far has defied visual discovery; most likely it is a white dwarf.

3. *Long-focus Photographic Astrometry*

The initial success of Bessel led me to the idea that photography with long-focus telescopes would furnish a promising technique for this problem. Long-focus photographic astrometry had been successfully introduced early in the twentieth century by Schlesinger for determining precise stellar parallaxes. While still at the McCormick Observatory at the University of Virginia, I decided more than four decades ago to introduce a limited programme to search for perturbations of nearby stars.

A full-fledged programme was begun in 1937 when I came to the Sproul Observatory of Swarthmore College, and put on the observing programme a number of stars within reach of the 61-cm refractor of 1093 cm focal length. Most stars on this programme are within 10 parsecs (33 light years); generally they are brighter than 11th magnitude, and for the most part north of the celestial equator. The path of the nearby star is generally measured on a background of three to six reference stars which for all practical purposes are very distant and 'fixed'. Several exposures of the order of one minute of time on two to four plates are taken on any one night, yielding a measured accuracy of about 1 micron or $0''{\cdot}02$ (probable error). By combining several nights, a higher accuracy of $0''{\cdot}010$ of even $0''{\cdot}005$ may be reached for a 'normal' point for any one year.

This programme, now under the direction of Sarah Lee Lippincott, is approaching four decades of its existence. Special intense observational coverage is aimed at for a limited number of the very nearest stars, since the chance for discovery appeared to be greater for these objects. Other observatories are making significant contributions in this field, primarily Alleghany and McCormick, and the U.S. Naval Observatory station, in Flagstaff, with its powerful 154-cm reflector.

4. *From Unseen to Seen Companions*

The classical unplanned photographic discovery of a perturbation for the faint red dwarf Ross 614 at a distance of 13·2 light years was made at the McCormick Observatory from a limited, only 25, number of plates extending over a prolonged parallax series, covering nine years (Reuyl 1936). A later study (Lippincott 1951) based on Sproul Observatory plates taken on 93 nights over the interval 1938-50, and McCormick plates taken on 27 nights over the interval 1927-37, led to a determination of the apparent orbit of Ross 614. Greatest apparent separation was predicted for 1955, at which time Walter Baade observed the companion, visually and photographically, with the Hale five-metre (200-inch) reflector (Lippincott 1955a, b). The period of this binary is 16·6 years; both components, Ross 614A and B are among the least massive stars known; as a matter of fact, a recent Sproul study based on plates taken on 252 nights over the interval 1938-72 (Lippincott and Hershey 1972) confirmed that Ross 614B is the least massive, known visible star, only six per cent of the Sun's mass. This value agrees with that predicted on theoretical grounds, it being the lowest possible mass for a *bona fide* main sequence star.

A study of the short period (0·278 days) eclipsing variable VW Cephei by John L. Hershey (1975) revealed a perturbation with a period of 30·4 years, which led to a predicted location, in which the third companion was seen by W. D. Heintz. Thus, at this time, four perturbations have led to subsequent visual discovery of the predicted companions; the classical cases of Sirius, Procyon, and the more recent cases of Ross 614 and VW Cephei.

An interesting example of a photographic study of a still unseen companion is the perturbation in the bright component of the visual binary BD+66°34=Mlb 377, which has a period of 15·9 years (Hershey 1973).

5. *Survey of Unseen Astrometric Companions*

A recent summary (van de Kamp 1975b) lists close to twenty perturbations, well established, for which thus far the companions remain unseen. The results depend on observations made at Allegheny, McCormick, Sproul, and U.S. Naval, Van Vleck and Yale Observatories. In all cases reasonable estimates of their masses and limiting values of their luminosities have been made.

The majority of the unseen companions seem to be faint dwarf M stars; a few appear to have masses less than the aforementioned theoretical value of six per cent of the Sun's mass. Unfortunately the lower the mass of the companion, the smaller the amplitude of the perturbation, which therefore is more subject and vulnerable to observational errors. The limitations of the method lie in the attainable accuracy of the photographic method. While an accuracy of $0''·010$ or better may be reached, one has to be on guard against systematic, slow or abrupt changes, inherent in the technique, which at times may introduce errors of several times $0''·01$. However, there are a number of well-established perturbations, fifteen having total amplitudes well over $0''·05$, and there need be no doubt about their reality.

6. *Planetary Companions*

Of particular interest in the potential discovery of planet-like companions. Thus far no other observational technique is able to detect planetary companions. Some astrometric evidence is available from a study made for the faint red dwarf Barnard's Star, the nearest star in the northern equatorial hemisphere. An intensive series of some four thousand plates has been taken with the Sproul 61-cm refractor on over one thousand nights covering the interval 1916-75. Measurements with the Grant two-coordinate machine reveal a perturbation with a total amplitude of only $0''\cdot02$, indicating the presence of a companion with a mass of our planet Jupiter, circling Barnard's Star with a period of 11·5 years (van de Kamp 1975a). There is additional, marginal, evidence for a second planet with a mass of 0·4 times Jupiter and a period of revolution of about two decades. It must be kept in mind that Barnard's Star is only 5·9 light years away and has only 15 per cent of the Sun's mass. The difficulty of finding planetary companions for more distant objects, and for massive primaries, is thus amply demonstrated.

7. *Total Mass of Unseen Companions*

The discoveries made thus far give for the first time some information of the cosmic mass in our neighbourhood, contained in unseen stars. In the perspective of the time scale of binary orbits, forty years of observational effort is just the beginning of this type of work, and the results are provisional for several reasons. Simple considerations show that incompleteness of discovery sets in even before the nearest star is reached. An attempt at making allowance for incompleteness and selection effects yields a *number density* of unseen companions of 0·21 per cubic parsec. The corresponding *mass density* is estimated at 0·063 solar masses per cubic parsec, a value virtually the same as that for visible stars (van de Kamp 1976). If this is an over-estimate it nevertheless emphasizes the incompleteness of discovery of the unseen objects. Dynamical studies based on statistics of stellar motions in the Galaxy indicate a total mass density of 0·15 solar masses per cubic parsec in a much wider solar neighbourhood (Oort 1965). All these values fit nicely, but should not be taken too seriously in view of observational limitations, selection and incompleteness effects. There is, of course, also the invisible mass contributed by interstellar material. All in all it behoves us not to draw premature conclusions but to rejoice that the orders of magnitude for the mass density of visible, unseen, and total population are reasonable and compatible. It will require decades of intensive observational efforts to make further progress in these fascinating aspects of the unseen universe.

References
Hershey, J. L. 1973. *Astron. J.* 78:935
Hershey, J. L. 1975. *Astron. J.* 80:662
van de Kamp, P. 1938. *Sigma Xi Quarterly* 26:103
van de Kamp, 1975a. *Astron. J.* 80:658
van de Kamp, P. 1975b. *Ann. Rev. of Astron. and Astrophys.* 13:295
van de Kamp, P. 1976. *Proceedings of Tercentenary Symposium, Herstmonceux* July, 1975.
Lippincott, S. L. 1951. *Astron. J.* 55:236
Lippincott, S. L. 1955a. *Sky & Telescope* 14:364
Lippincott, S. L. 1955b. *Astron. J.* 60: 379
Lippincott, S. L. and Hershey, J. L. 1972. *Astron. J.* 77:679
Nörlund, N. E. 1905. *Astron. Nachr.* 170:9
Oort, J. H. 1965. *Galactic Structure.* Univ. of Chicago Press: 472
Reuyl, D. 1936. *Astron. J.* 45:133
Seeliger, H. 1914. *Astron Nachr.* 199:273
Strand, K. Aa. 1942. *Astron. J.* 49:165

Ice in Space

M. COHEN

Anyone who has seen the dark lane in the sky which is the Great Rift of the Milky Way through Cygnus and Aquila, or has looked at the Orion Nebula through a telescope, can easily believe that our Galaxy is full of dark, dusty, obscuring clouds. This dust exists in the form of abundant small grains, with sizes typically between one and a few hundred millionths of an inch. The particles occur in three different types of environment: in the largely empty spaces between the stars – the 'interstellar medium'; in the immediate vicinity of certain classes of star – in the so-called 'circumstellar shells'; and in the dense, dark-cloud complexes in which anything from a few hundred to a few hundred thousand times the mass of our Sun are clumped together in the form of gas and dust, and in which new generations of stars are even now being born.

Dust plays a vital role in the cycle of stellar birth and death. It enables gas clouds which are collapsing to become dense enough at their centres to create new stars. It collects into disks about the newly formed stars and provides the material for planetary systems. It is produced at the end of stellar life as a sort of smoke that returns the heavy elements cooked up inside the stars during nuclear burning to the interstellar medium. Agglomerations of interstellar gas and dust yield new dark clouds, and the cycle is complete.

Another important process that occurs in the dark clouds is the production of molecules. In such clouds, where the temperatures are very low (at most some tens of degrees above absolute zero) and the gas densities are quite high, the surfaces of dust grains

ICE IN SPACE

provide both an environment in which atoms may come together and join into molecules, and shielding from destructive stellar ultra-violet photons (that would break up, or 'dissociate' the molecules). What are these molecules that we expect to be built up in the clouds, and what evidence do we have for their existence?

The relevant theoretical calculations involve both the physics and chemistry of molecular formation and dissociation, and must take account of the observed relative amounts of elements that we find in the Universe. In other words, there is a reservoir of hydrogen atoms available as basic building blocks for molecules, with appreciable quantities of carbon, nitrogen, oxygen, and lesser amounts of silicon, magnesium, and iron. Even this simple kit is sufficient to assemble a wealth of organic and inorganic molecules (with, and without, carbon, respectively). The most abundant simple molecules would be water, ammonia, and methane, and these could occur either as gases (vapours) or in frozen form as ices. Carbon monoxide gas would be plentiful also.

A molecule may be thought of as a set of beads (atoms) attached by springs (the atomic bonds). As such, a molecule may be set quivering in various ways – the springs can stretch and compress, bend, or twist, or undergo some combination of these activities. Each of these actions would cause the molecule to have a particular type of energy which in the course of time it would release so as to return to its preferred, lowest energy state (just as a ball will prefer to roll down a slope rather than rolling up, which increases its energy, or staying fixed, which maintains a positive energy). Now the return to this lowest state of energy involves the release of a quite specific and determinate amount of energy dictated by the types of atom the molecule is made of, and the type of vibration involved (stretching, bending, or rotation). These determinate amounts of energy are liberated as radiation at specific frequencies and a single molecule produces a recognizable 'spectrum', or pattern of frequencies. Additionally, the different types of quivering produce frequencies in different parts of the electromagnetic spectrum. Pure rotation (twists of the springs) yields 'microwave' frequencies (radio waves with lengths of millimetres and centimetres). Vibrations are seen in the infra-red (waves some five to twenty times longer than the reddest that the human eye can perceive). In this manner, it is possible for an astronomer to play detective by studying a set of fingerprints (a spectrum) and recognizing both the specific molecule responsible for the prints and the type of motion it was undergoing when it emitted the radiation. The pattern recognition is not always this simple, however, for if rotation accompanies another vibration, then the frequencies are blurred – we do not see simple lines in the spectrum, but rather broad bands. Secondly, there is no reason why molecules should only be seen releasing energy (the ball rolling down the slope). Some mechanism has to provide that energy (a foot perhaps?) and it is therefore possible to see the fingerprints either as bright bands (emission = energy released) or as dark bands (absorption = energy sucked in).

Radio astronomers have detected and identified a large number of organic and inorganic molecules by their bright and dark rotational fingerprints. Indeed, our Galaxy has a substantial fraction of its total mass in the form of enormously extended clouds principally of hydrogen and carbon monoxide. In addition to clouds, very small, localized sources of hydroxyl, water vapour, ammonia, and formaldehyde are known, to list but a few from the chemist's shop. Now if there is so much water vapour, and temperatures in space are low, where is the water ice that theoretically should be present?

Ice in the laboratory produces a strong absorption fingerprint centred at about 3·1 microns in the infra-red (1 micron = a thousandth of a millimetre) and this is the best frequency at which one should search. The first problem was one of technology – the special detectors that infra-red astronomy needed were only developed in the early 1960s. Looking for molecular features requires that one divides the received radiation into many small pieces to define the presence and shape of suspected absorption or emission features. But the new detectors lacked the sensitivity to do more than detect broad pieces of spectrum so spectroscopy was delayed some years. By the late 1960s astronomers were using

ICE IN SPACE

better equipment, both on high altitude balloons and on the ground, to find water vapour and seek ice. It was known that grains and simple molecules were located in the interstellar medium; so why not ice too? What was needed was some long path length with a large, known amount of dusty obscuration to ensure that ice grains too would be expected along this line of sight in appreciable quantities. A highly obscured star in Cygnus was chosen, which is known to suffer ten magnitudes of interstellar dimming. That is to say – if not for the intervening interstellar medium, this star would appear as of second magnitude and not twelfth!

Despite this apparently favourable environment, no ice absorption was detected, and even as recently as 1975, working with far more sensitive equipment, no ice was seen against this star. However, in 1972, a feature matching laboratory data on ice absorption reasonably well was found – against a bright infra-red source inside the Orion Nebula, which source is so heavily obscured as to be optically invisible. This infra-red source was thus used as a probe of the intervening dark clouds associated with the Orion region, and was seen shining through an appreciable quantity of ice as well as gas and dusty material. The estimated amount of obscuration in this direction is almost unbelievably high – about 50 magnitudes! If this amount of obscuration could be placed between our Sun and the Earth, not even a deep exposure with the Palomar 200-inch reflector would record the Sun on a photograph as more than the tiniest grain! Virtually all of this 50 magnitudes occurs inside the dense Orion clouds and not along the approximately 1500 light years that separate us from the Orion Nebula.

In the last two years a breakthrough in technology has made extremely sensitive detectors available for infra-red astronomy, and these have yielded ice features in several more infra-red sources. With one exception, all of these sources are optically invisible, indicating enormously thick obscuring clouds, which are believed to be local to the infra-red sources. Hence, even though ice is only a very minor constituent of the dust in space, enough of it intervenes between us and these sources that a recognizable absorption feature is caused. The true nature of these infra-red objects is difficult to evaluate – they could be quite massive stars that have formed inside dark clouds and are now ageing and maturing but are still shrouded in their dusty curtains. The lack of light available to us means that all direct (optical) spectroscopic tests to determine what the infra-red sources are cannot be applied. It would appear, however, that the infra-red emission is due to radiation from heated dust particles in the vicinity of the objects, but just what sort of stars might be heating the grains, and even whether they are young or old objects, can only be conjectured at present.

One exception is outstanding – there exists one optically known object against which ice has been detected. One of my particular interests is star formation. The T Tauri variable stars are a most interesting family of optical variables that are very young, typically about a million years (our Sun is 6000 million years old), and which do show large excesses of infra-red energy, most of which is due to heated dust grains in the circumstellar regions. When the new detectors became available I was most interested to look for molecules near young stars. The programme included some 21 stars, mostly T Tauris, with a sprinkling of more massive stars and some enigmatic objects. Let me explain what I mean by 'enigmatic' here. Virtually all stars when examined with a spectroscope reveal a bright continuous rainbow of light crossed by narrow dark bands. These bands are due to atomic fingerprints principally, although the fuzziest are again caused by molecules. Once the atoms and molecules have been identified it is possible to decide upon the temperature and evolutionary status of the star. Suppose now that one finds a star that has no dark fingerprints at all. What temperature is it? Is it young or old? You see the problem. I observed just such a star, that is believed to be very young by virtue of its association with other T Tauri stars in the Taurus dark-cloud 'nursery', some 500 light years away. It is HL Tauri, of visual magnitude 15. It is very bright in the infra-red in a manner that suggests it is embedded in a thick dusty shell, and it reveals no optical fingerprints at all! In this case, I believe that dense dust surrounding this very young star is responsible for veiling the spectrum fingerprints. On the basis of other, independent, evidence others have argued too for HL Tauri's extreme youth, perhaps a mere 300,000 years.

I observed HL Tauri in late 1974, and the analysis of these observations revealed a deep fingerprint where the ice band should be. It was with great excitement that I returned to the telescope in early 1975 to try to confirm the earlier findings. For five consecutive nights in a seven-night observing spell our mountain was shrouded in clouds and snow. On the sixth night the sky cleared, and HL Tauri was re-observed at last. The deep absorption repeated itself. The next step was to compare the observed feature with that known for ice in the laboratory. The match was excellent! This match proved that water ice had been detected for the first time in the direction of a visible star, and it yielded information on the likely size of the ice grains (no bigger than some 20 millionths of an inch) and the quantity of ice that intervenes between HL Tauri and the Earth.

Where is this ice seen against HL Tauri? It cannot be in interstellar space, for ice is not detected against the other T Tauri stars in the Taurus dark clouds, even though they are all brighter in the infra-red, which would make recognition of a fingerprint much easier than for HL Tauri. The ice must be in the dusty circumstellar region where veiling of the optical spectrum of HL Tauri occurs. And how much ice is there? By comparison with the Orion Nebula source, Orion has three times the quantity of ice that HL Tauri has, but more than ten times the optical obscuring material of HL Tauri. In other words, the environment of the very young star is exceptionally rich in ice grains, given the small visual obscuration. Indeed a most interesting environment!

Two rather peculiar young stars that have each suddenly brightened by six magnitudes optically in the past, and then remained relatively constant in brightness were also observed in my programme. No ice is seen, but these stars, and only these stars, show fingerprints of water vapour.

Collecting the facts together we can say that where extremely dark obscuring clouds are found hiding infra-red sources one may sometimes find ice. Among the young stars, however, circumstellar ice is only found around the very youngest stars, and then it is unusually richly present. The fate of ice close to a young star is to be vaporized as the star heats up and matures, melting the ice grains. Perhaps the water vapour around the two 'flaring' young stars mentioned above is a left-over of an unusually rapid melting of original ice grains. But this is where facts end and speculation begins, and this is where our story must end for now!

The Dusty Sky: H II Regions

DAVID ALLEN

It is characteristic of an old science such as astronomy that new ideas are only slowly absorbed into its fabric. Innovations of technique equally are shrugged aside by many of the more orthodox observers weaned on traditional methods. To appreciate this tendency one has only to note the unseemly delay between Jansky's discovery of radio emission from the Galactic Centre and the first professional radio observations. This delay amounted to a decade and a half, a figure which may be seen in context by considering how much radio astronomy has taught us during the subsequent three decades.

So it is with other portions of the electromagnetic spectrum. The last decade has witnessed the opening up of the spectrum to astronomical investigation. Balloons, rockets, and satellites have carried above the impeding atmosphere detectors of X-rays and γ-rays, ultra-violet and infra-red radiation, detectors which in many cases have been long in use in physics laboratories. This widening of our horizons beyond the parochial atmospheric windows accessible from the ground is making a great impact on astronomy, but it is an impact which was long overdue.

Take infra-red astronomy as an example: it is a century and three-quarters since Sir William Herschel discovered that the Sun emits energy beyond the red end of the visible spectrum. With equipment no more sophisticated than a prism and a thermometer Herschel began to open up that great section of the electromagnetic spectrum which separates the optical and radio domains. One hundred years ago the fourth Earl of Rosse was using infrared techniques to measure the range of temperatures on the surface of the Moon. Forty years ago Pettit and Nicholson employed the same techniques to examine the temperature variations of Mira variables throughout their ponderous cycles. But you can open almost any modern treatise on elementary astronomy and read that from the surface of the Earth we can make astronomical observations at *only* optical and radio wavelengths. In the last ten years infra-red astronomers using normal telescopes at established observatories have precipitated an era of discovery: their data have bruised many of our concepts and forced a serious re-appraisal of our theories of most celestial objects. And still the elementary texts make no mention of the subject.

By observing at infra-red wavelengths we examine objects cooler than normal stars. The cooler an object, the deeper into the infra-red is its radiation emitted. Anything cooler than about 1200°K cannot be recorded at optical wavelengths and radiates solely in the infra-red. Stars at such low temperatures are not known; instead our infra-red detectors record principally the radiation from dust grains. This statement summarizes the essence of infra-red astronomy: over the last decade we have demonstrated the existence of vast clouds of dust in space, many of them in the most unexpected of places. Most types of celestial object, from the comets that make occasional pilgrimages into our Solar System to the farthest quasars, have associated with them clouds of dust. This dust causes the sources to be brighter than expected at infra-red wavelengths.

The dust is heated by stellar radiation. Just as the Earth absorbs sunlight and settles to an equilibrium temperature, so each grain of dust absorbs energy from its nearest star and adopts an appropriate temperature. The energy it absorbs from the star is re-radiated in the infra-red Of course, a tiny grain of dust can absorb only a miniscule amount of radiation, and can radiate correspondingly little. We can detect the infra- red radiation from only enormous numbers of these grains, forming clouds with masses much larger than that of our entire Solar System. Such clouds, it seems, are common. As an example of the occurrence of dust in celestial ojects I have chosen H II regions.

THE DUSTY SKY: H II REGIONS

The presence of dust clouds in and around H II regions is well known. We need look no further than to the constellation of Orion, which on long-exposure photographs takes on a blotchy black and white complexion. The bright portions are the clouds of ionized gas – the H II regions – which glow by the emission lines of their hydrogen and other ions. The darker areas are not simply gaps between these H II regions. Some, especially those with sharp edges, are foreground dust clouds so dense that no light passes through. The Horsehead Nebula in Orion is one of the best-known such clouds. The Horsehead, however, does not radiate in the infra-red (or if it does it is too feeble to have been detected). This is because there are no nearby stars to heat the dust. But wherever there are stars to heat the dust the gas is ionized. In short our infra-red telescopes record the dust inside or very near the H II regions.

Most of this dust is at quite low temperatures – typically 100°K. Its radiation is therefore emitted at the very long wavelengths around 50 microns (the size of the grains on a photograph, about one hundred times the wavelength of visible light). At 50 μm the Earth's atmosphere is opaque; most infra-red observations of H II regions have therefore been made from rockets, balloons, or high-flying aircraft. Such observations are difficult to perform and extremely costly. Our knowledge of the dust clouds in H II regions is but slowly being prised from the reluctant skies. The longest wavelength at which we can readily observe from the Earth's surface is 25 μm. Dust at 100°K radiates only weakly at this wavelength, but this is compensated by the much larger telescopes available. The first infra-red observations of H II regions were made with ground-based telescopes, and the most detailed maps have also been produced in this way, principally by Eric Becklin, Gerry Neugebauer, and Gareth Wynn-Williams using the Palomar 200-inch telescope.

At 20-25 μm the brightest H II region is Messier 17, the Omega Nebula. The energy output of this object in the infra-red is brighter than a million suns. This implies that there are stars in M 17 whose visible output at least equals this figure. We cannot see these stars because of the intervening dust.

At longer wavelengths many more H II regions radiate strongly. Indeed, Messier 17 ranks only third in the field, being surpassed by Sagittarius B and, brightest of all, the Orion Nebula Messier 42. Some of the H II regions recorded at this wavelength are optically invisible and do not record on photographs. They are, however, strong radio sources. To account for the absence of visible nebulæ we need not look far: dust is again the answer. The dust clouds which obscure these H II regions lie in the spiral arms of the Galaxy. They are the dust lanes which would give our Galaxy the appearance of the famous Sombrero Hat (NGC 4594; M 104) if we could view it edge-on from a distance. This interstellar smog obliterates most of the H II regions in the constellations of Sagittarius, Scorpio, and Centaurus, including Sgr B, a giant cloud near the centre of our Galaxy.

Messier 42 was the first H II region found to emit in the infra-red. In 1967 Doug Kleinmann and Frank Low of the University of Arizona discovered a very bright source about 1' arc south of the Trapezium cluster. The Kleinmann-Low nebula, as it became known, is the centre of the infra-red emission of M 42, a cloud of cold dust that extends over most of the constellation of Orion and in a few regions is warmed by nearby stars so as to emit infra-red radiation. Fainter infra-red sources coincide with Messier 43, where the dust is heated by NU Ori, and the Trapezium group itself. Another region of infra-red emission lies a few minutes of arc north of the Trapezium in a dark portion of the nebulosity. This infra-red source was discovered as recently as December 1973.

Because M 42 is the nearest and brightest major H II region, and one of the youngest to boot, it is the best studied. By combining optical, infra-red, and microwave data we have been able to construct a picture of this object which is probably applicable at least in part to many other H II regions. The picture is a surprising one.

The most surprising feature is the size and density of the nebula.

THE DUSTY SKY: H II REGIONS

Far from being a smallish, tenuous cloud of ionized hydrogen it is a large, dense conglomerate of gas and dust many hundred times more massive than the Sun. Throughout most of the cloud the hydrogen is neutral, not ionized. Indeed it would be invisible but for the Trapezium stars which ionize the outer regions and cause them to glow. The Trapezium stars have formed only recently near our edge of the cloud and have been burning their way out ever since. They have nearly succeeded: there remains an arm of dark neutral gas and dust to the east which curls round nearly in front of them and forms the dark bay in the nebula.

Deeper inside the neutral cloud all is hidden. We cannot know if, for example, there are other Trapezium-like stars burning their ways out of the far side. We do, however, find evidence that some sources exist somewhere in the midst of the cloud. In the vicinity of the Kleinmann-Low nebula is a cluster of infra-red point sources. The brightest of these sources is known as the Becklin-Neugebauer source, after its discoverers.

It is possible that the Becklin-Neugebauer source and its satellites are a cluster of extremely luminous stars much brighter than the Trapezium cluster. On this theory they are extinguished by the dust and thus rendered invisible at optical wavelengths. The infrared radiation, on the other hand, penetrates dust clouds with greater facility, allowing us to record their presence. An alternative, and perhaps slightly more attractive, hypothesis makes the infra-red sources protostars – knots of gas and dust which are in the process of contracting from the cloud to form stars, but which have not yet reached the nuclear burning phase.

For the sources in M 42 it is difficult to decide between these two hypotheses. Since similar infra-red sources are turning up in many other H II regions and dark neutral clouds it is likely that we will soon be able to eliminate one or other hypothesis for some, and by inference perhaps for all such objects.

Before the advent of infra-red astronomy we thought we understood H II regions. We imagined them to contain small amounts of dust, and we knew that in a few there were stars hidden from direct view by that dust. But it never occurred to us that H II regions could emit most of their energy at infra-red wavelengths. Nor did we suspect the existence of objects like Becklin and Neugebauer's. The greatest lesson for us to learn in astronomy is that we know and understand less than we think.

Viking to Mars

PATRICK MOORE

Almost exactly seven years after the first manned landing on the Moon, America's unmanned Viking 1 touched down upon the surface of Mars and began sending back information. The triumph was spectacular by any standards, and it has been said, with truth, that Viking ranks as the 'cleverest machine' ever built.

Of course the most fascinating part of the whole programme was the search for life. Unfortunately, at the time when this article is being written (27 Sept. 1976) the results are inconclusive, and scientists at the Jet Propulsion Laboratory are frankly puzzled. By the time that the *Yearbook* appears in print we may know much more, but for the moment it may be wisest to concentrate upon other aspects of Viking – and to give some idea of how dramatically our ideas about Mars have changed during the past couple of decades.

As a start, let us list all the Mars probes which have been dispatched to date. They are as follows:

Mars 1 (U.S.S.R.). Intended flyby. Launched 1 November 1962. Apparently it was put into the correct path, but contact with it was lost at a distance of 65,900,000 miles from Earth and was never regained, so that nobody knows what happened to it. Presumably it is still orbiting the Sun.

Mariner 3 (U.S.A.). Intended flyby. Launched 5 November 1964. Total failure. The probe itself is in solar orbit, but shroud failure prevented the Mars flyby.

Mariner 4 (U.S.A.). Mars flyby. Launched 28 November 1964, and on 14 July 1965 it by-passed Mars at a distance of 6,118 miles, sending back 21 pictures as well as invaluable information about the planet's atmosphere. Craters were recorded, and it was found that the atmospheric pressure was unexpectedly low – far less than the previously-accepted figure of 85 millibars or thereabouts. Also, the atmosphere proved to be made up chiefly of carbon dioxide rather than nitrogen, and the old theory due to Johnstone Stoney – that the white polar caps were of solid CO_2 rather than water ice – came back into favour. After its encounter with Mars, Mariner 4 continued in solar orbit. Signals from it finally ceased on 20 December 1967.

Mariner 6 (U.S.A.). Launched 24 February 1969, and passed over the Martian equatorial region on 31 July of the same year at a distance of 2,120 miles. Its 75 pictures confirmed the existence of craters, and it also supported the new values for the density and composition of the atmosphere.

Mariner 7 (U.S.A.). Launched 27 May 1969, and passed over the southern hemisphere of Mars on 4 August, sending back 126 pictures, of which 33 were of the south polar region. Both it and its twin are now in orbit round the Sun. After their flights, Mars was generally dismissed as a rather uninteresting world – a kind of cratered waste. We now know that by sheer bad luck both probes missed the most important areas in their surface photography.

Mariner 8 (U.S.A.). Launched 8 May 1971. Complete failure; the second stage failed to ignite, and the probe fell unceremoniously into the sea. Officially it is still referred to as Mariner H. It had been intended to go into a closed path round Mars.

Mars 2 (U.S.S.R.). Launched 19 May 1971; intended orbiter and soft-lander. It reached Mars on 27 November, and entered a closed orbit which presumably still carries it round the planet in a period of 18 hours, with a distance ranging between 1,530 miles and 15,250 miles from the surface. It ejected a capsule carrying a Soviet pennant, which seems to have landed safely, though whether the achievement was scientifically notable is a matter of opinion!

Mars 3 (U.S.S.R.). Launched 28 May 1971; another intended orbiter and soft-lander. The orbiter is still circling Mars in a period of 11 days; the path is very eccentric, with the distance ranging between 970 miles and 133,000 miles. The ejected capsule came down at latitude 45°S., longitude 158°W., between Electris and Phæthontis. The package was activated $1\frac{1}{2}$ minutes after touch-down (13h 50m 35s UT) but after twenty seconds it went dead, and no further signals were received. The reason for the failure is unknown, but in view of what Viking has now told us it may well be that the lander came down on uneven ground, so that it was tilted by an unacceptable amount. No doubt some future explorers will go and search for the remains.

Mariner 9 (U.S.A.). Launched 30 May 1971; the twin of the ill-fated Mariner 8. On 13 November 1971 it entered Mars orbit (distance ranging from 1,025 miles to 10,500 miles), and between then and 27 October 1972 it returned 7,329 high-quality pictures. Before Viking, it provided us with most of our detailed information about the planet, and the first high-resolution maps were drawn up.

Mars 4 (U.S.S.R.). Launched 21 July 1973, the first of the four-vehicle 'fleet'. It approached Mars on 10 February 1974, but missed the planet – presumably it had been destined to go into a closed orbit – and must be classed as a failure.

Mars 5 (U.S.S.R.). Launched 25 July 1973, and went into Mars orbit on 12 February 1974, but seems to have sent back little positive information.

Mars 6 (U.S.S.R.). Launched 5 August 1973; intended orbiter and soft-lander. It arrived on 12 March 1974, but signals from the lander ceased during the descent, and contact was never re-established.

Mars 7 (U.S.S.R.). Launched 9 August 1973; arrived in the vicinity of Mars on 9 March 1974, but it too failed to go into a closed orbit. All in all, the Mars fleet was signally unsuccessful, though a few pictures were obtained (far inferior to those from Mariner 9).

Viking 1 (U.S.A.). Launched 20 August 1975; entered Mars orbit on 19 June 1976, and the lander came down in Chryse on 20 July.

Viking 2 (U.S.A.). Launched 9 September 1975, and entered Mars orbit on 7 August 1976, with a view to sending its lander down to the region of Cydonia. (It actually came down Utopia.)

This, then, is the story to date, and it is notable that all the real successes have been American; the Russians have had no luck at all, which is rather strange in view of their triumph in obtaining pictures from the far more hostile Venus. It must also be added that Mariner 9 and Viking 1 have also obtained excellent pictures of Phobos, the senior of Mars' two midget satellites, which is shaped rather like a potato, and is heavily cratered; Mariner 9 also sent back pictures of the even smaller Deimos. As expected, both satellites have synchronous rotation; that is to say, each keeps the same hemisphere turned toward Mars all the time. They may or may not be captured asteroids; at any rate, they are totally unlike our own massive Moon.

The famous Martian canals were among the earliest casualties of the Space Age. As most people know, they were first described in detail by the Italian astronomer G. V. Schiaparelli in 1877, though in fact they had been drawn earlier; Percival Lowell, who founded the celebrated observatory at Flagstaff principally to study Mars, was utterly convinced that they were artificial, and he and his colleagues covered the planet with a network of geometrical lines. Other astronomers were sceptical, and regarded the canals as optical illusions. This has proved to be the correct explanation; there is nothing even remotely artificial-looking anywhere on Mars, and the canal *réseau* simply does not exist. (Let it be added that Lowell carried out a great deal of valuable work in other branches of astronomy; for instance, it was his calculations which led to the tracking-down of the ninth planet, Pluto. It would be most unfair to remember him only because of his erroneous theories about Mars.)

Before Mariner 4 it was widely supposed that the dark areas were due to plant life – or at least to organic material of some kind. One argument in favour of this attractive idea was put forward by E. J. Öpik, of Armagh Observatory. Öpik pointed out that there is plenty of dust in the Martian atmosphere, and he maintained that the dark areas would soon be covered up unless they had regenerative power – that is to say, the ability to push the dust out of the way. This seemed reasonable, but it too has proved to be wrong. Neither are the dark areas depressed basins – some of them, such as the Syrtis Major, are elevated plateaux.

Then there was the question of the nature of the polar caps, which undoubtedly wax and wane with the Martian seasons. They had been regarded as snowy or icy, even though it was recognized that the depth must be very slight; otherwise, the shrinking of a cap during spring and early summer would release large quantities of water vapour into the atmosphere, which was in flat contradiction to the spectroscopic evidence. Johnstone Stoney's theory of solid carbon dioxide caps was confirmed – at least in the main; and this leads us on to the 'wave of darkening' which had been widely accepted by many observers.

It had been maintained that when a cap began to shrink, releasing moisture, the vegetation tracts near the poles began to darken and harden, as though the organisms were being revived by the arrival of water vapour. This wave of darkening continued through to the equator, and the seasonal cycle was repeated year after year. Personally I was always dubious, chiefly because I had never been able to follow the process myself even with the help of powerful telescopes, and now that we are sure that the dark areas are not vegetation-tracts the phenomenon would be very hard to explain – all of which underlines the difficulty of making accurate observations of a small world which never approaches us much within 35 million miles.

Mariner 9 sent back pictures of giant volcanoes, one of which, Olympus Mons – formerly known as Nix Olympica – towers to over 15 miles above the general ground level, and is crowned by a 40-mile caldera. Olympus Mons is a shield volcano, similar in type to those of Hawaii, but on a larger scale. It is not unique; there are other volcanoes of comparable height, and one of them (Arsia Mons) has an even larger summit caldera. Associated with the volcanoes are what seem to be drainage systems; there are tremendous canyons, and there are also many winding features which look so like dry river beds that it is hard to believe them to be anything else. No appreciable liquid water can exist on Mars today, because the atmospheric pressure is too low (below 10 millibars everywhere), but we are forced to the conclusion that there must have been water in the fairly recent geological past, so that the atmosphere must also have been much denser.

One explanation is associated with the changing axial tilt of Mars. At present the angle of inclination is about the same as ours (between 23 and 24 degrees), but over long periods it can range between 35 degrees and only 13 degrees. When the tilt is at its maximum, it is suggested that the poles receive a greater amount of total heat, so that the volatiles in the white caps are released into the atmosphere to produce a temporary thickening. Whether this explanation is correct remains to be seen. According to a minority view there may be occasional periods of intense volcanic activity, so that gases – and water vapour – are expelled from below the Martian crust. Of course, it is always possible that both processes may have played a rôle.

Much of Mars is reddish-ochre in hue, and Lowell compared its colour with that of the Painted Desert of Arizona. One of these desert areas is Chryse, south of the famous dark feature once known as the Mare Acidalium and now re-christened Acidalium

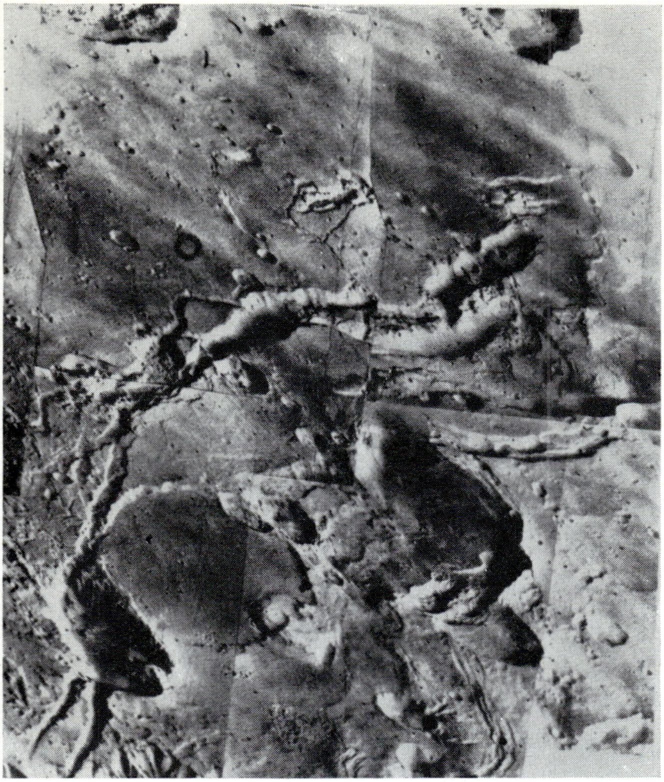

Figure 2. Fault-zones in Cydonia, photographed from Viking 1.

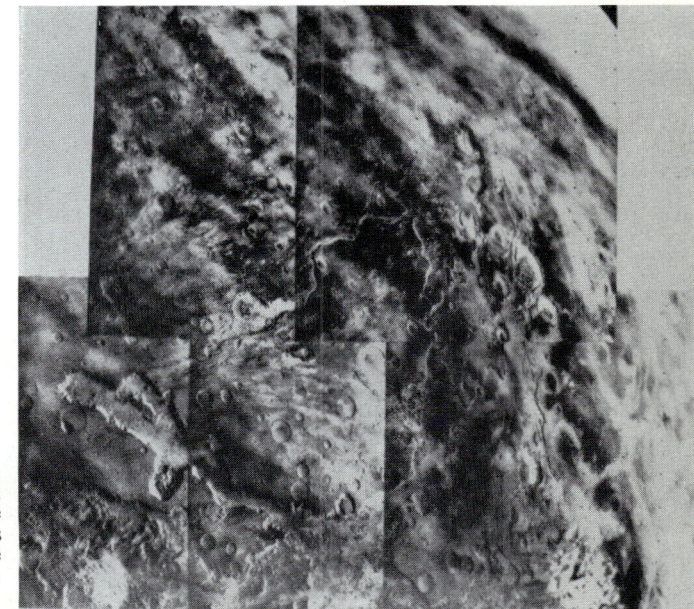

Figure 3. Chryse, photographed from Viking 1 at a distance of 20,000 miles. Clouds can be seen to the lower right.

Figure 4. The 11-mile crater Yuty, in Chryse.

Figure 5. First photograph of the foot of the Viking lander after arrival.

Figure 6. Scene from the lander, 28 July 1976, p.m. Mars time.

Figure 7. Trench dug by the arm of Viking 1, 3 in. wide, 2 in. deep and 6 in. long. This was obtained on the 8th sol after arrival.

Figure 8. The rock-strewn surface of Chryse, from the lander of Viking 1.

All photographs are reproduced by courtesy of N.A.S.A.

Planitia. Chryse was expected to be as smooth as most areas of the planet, and was of interest because it lay not too far from the end of the huge canyon now called the Vallis Marineris – part of the drainage system associated with Olympus Mons and the other great volcanoes of the Tharsis ridge. Obviously, then, the first task was to survey Chryse as thoroughly as possible. Preliminary work was carried out by Earth-based radar, using the radio telescope which has been built in the natural bowl at Arecibo, in Puerto Rico, and faint feelings of alarm were roused, because Chryse seemed to be rather less smooth than had been hoped. When Viking went into a path round Mars and sent back detailed pictures of the area, it became clear that the original landing site was unacceptable. A second was selected, some way to the north-west, and then a third, still further west but still within the region of Chryse. It was here that the landing was finally made, on 20 July 1976.

The main risk was that if the lander came down on top of a sizeable rock it would certainly be put out of commission, and a tilt of 19 degrees was the maximum permissible. Also, the landing manœuvre had to be entirely automatic; once started it could neither be stopped nor modified, and rocks big enough to be dangerous were still too small to be revealed either by radar or by the Viking surveys. Luck was bound to play a part – and it did. We now know that the lander touched down a mere 25 feet from a rock which was 10 feet in diameter and 3 feet high. When the first signals came through from the Martian surface, there was understandable jubilation among the scientists who had planned the experiment with such tremendous care.

Work began almost at once, and before long the first 'weather reports' came through. It was fine and sunny at Chryse, but also very cold, with a temperature ranging between -20 degrees Fahrenheit, at 2 p.m. Martian time, down to at least -120 degrees Fahrenheit. The sky was cloudless, though some dusty layers could be made out, and the dominant colour was pink rather than blue, because of the fine dust suspended in the atmosphere. Winds were light, reaching a maximum of around 14 m.p.h. at 10 a.m.

and dropping to only about 4 m.p.h. in the evening (Fig. 1). Note that it is quite permissible to give times according to our familiar units, because a Martian day – now known officially as a 'sol' – is only a little over half an hour longer than ours.

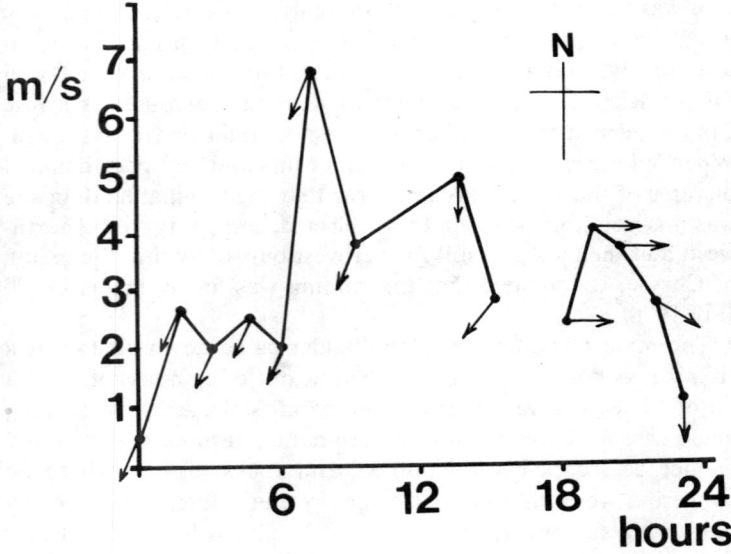

Figure 1. Windspeeds on Mars, measured over 1 sol. The speeds are given in metres per second.

Analysis of the atmosphere showed that, as fully expected, most of it was made up of carbon dioxide. In fact, this gas accounted for about 95 per cent, with 2 to 3 per cent of nitrogen, 1 to 2 per cent of argon, and a mere 0·3 per cent of oxygen, with no more than a trace of water vapour. When the lander arrived the pressure was about 7 millibars, which is not very much; and over the next few sols it dropped appreciably. The reason was straightforward enough. Winter was coming to the southern hemisphere of Mars, and the southern cap was forming, so that carbon dioxide was being withdrawn from the atmosphere and causing a reduction in pressure. (Chryse, incidentally, lies some way north of the Martian equator.)

One of the most important parts of the programme involved sending out an 'arm' to scoop up some surface material and draw it back into the body of the space-craft for analysis. The search for possible life depended entirely upon this, and there were misgivings when the arm showed signs of giving trouble; fortunately it worked in the end, and adequate supplies of 'soil' were collected. It was found that the composition of the material was not unlike that of the lunar maria, with appreciably less silicon than in most terrestrial basalts.

The pictures which came back soon after Viking's arrival were spectacular by any standards, and were of amazingly good quality (See *Figures 2-8*). Rocks were strewn everywhere, and, rather surprisingly, many of them were sharp instead of being rounded and worn away by dust erosion. This was puzzling, and as yet the solution is not entirely clear. It has been suggested that winds have blown away the dust following the great storms of 1971-2, exposing the rocks beneath; it may also be that there has been a considerable amount of 'fatigue stress'. At any rate, the surface has proved to be quite firm enough to support the weight of a space-craft such as Viking. Viking 2 came down in the same kind of landscape.

By the time that the lander had been transmitting for a few sols it had told us more about Mars than would have seemed even remotely possible a few years ago. The picture that emerges is of a chilly, inhospitable world, rock-strewn and cratered, with volcanoes which may or may not be totally extinct. Yet Mars is much less unfriendly than Venus, with its dense, choking atmosphere, its hopelessly high surface temperature and its clouds of corrosive sulphuric acid; and there can be little doubt that it will be the first target for manned exploration following the setting-up of permanent bases on the Moon. It is still far too early to forecast just when the first expedition will set out; certainly we will need better vehicles than can be built at the moment, but at least a journey to Mars now seems less fantastic than a flight to the

Moon would have seemed before the war. Viking has been an essential link in the programme, and nobody with a real interest in the exploration of the Solar System is likely to forget that great moment when we on Earth received the first messages from the Red Planet.

Recent Advances in Astronomy

PATRICK MOORE

It is seldom nowadays that a year passes by without some spectacular development in astronomy, but very often the most exciting events are such that few people can see the results directly. However, 1976 has produced something really prominent: West's Comet, which may not have been comparable with the brilliant visitors of the nineteenth century, but which made a brave show for some time. It was, in fact, the brightest comet for years, and certainly far superior to the famous (or notorious!) Kohoutek's Comet which proved to be so disappointing visually. West's Comet was also interesting, inasmuch as it showed signs of disintegration after it had passed through perihelion. Its period is immensely long, but even when it eventually returns it will certainly not be as brilliant as it was in 1976.

Another interesting comet was discovered on 25 February 1976 by Schuster, in Chile. It never became bright, because it was always remote; its perihelion distance was 6·89 astronomical units or 639 million miles – well beyond the orbit of Jupiter, and setting up a new cometary record.

Meteorites are not associated with comets (at least according to majority opinion), and major falls are rare. One occurred on 8 March 1976, near Kirin in north-east China. This was seen as a brilliant fireball, which exploded and dropped stony meteorites over a wide area; the largest fragment collected weighed 3900 lb (1770 kg), beating the 'aerolite record' previously held by the Norton-Furnas meteorite of 1948 – which weighs a mere 1078 kg, and is to be seen at the Institute of Meteoritics in Albuquerque. According to Chinese sources, over a hundred fragments of the Kirin Aerolite were collected.

Of course the main 'space event' of 1976 was the Viking programme, described separately, but the Russians also have been active. Cosmonauts Boris Volynov and Vitaly Zhobolov made a prolonged flight in the space-station Salyut 5, and it was generally supposed that they might try to break the endurance record set by the last crew of Skylab, but in fact they did not do so, and returned safely to Earth after less than fifty days. More Cosmos satellites were launched (the grand total is now over 850), and in August 1976 there came a new lunar probe, Luna 24.

The probe was dispatched on 9 August, and was put into a circum-lunar orbit. On 18 August, at 9.36 a.m. Moscow time, it came gently down in the hitherto-unexplored Mare Crisium, at latitude $+12°$ 45', longitude $62°$ 12' E – not far from the touch-down-point of the unsuccessful Luna 23 in November 1974. After landing, Luna 24 deployed its soil-scooping device, drilled down to a depth of two metres, and collected samples. At 8.25 a.m. Moscow time on 19 August it took off again, and landed safely 200 km south-east of Surgut, in Siberia, at 8.55 p.m. Moscow time on 22 August. The whole experiment seems to have been carried through without a hitch, and it is really important to obtain samples from this region of the Moon. At the time when this article is being written (25 August 1976) analysis of the Mare Crisium material is being undertaken, and the results will no doubt have been published by the time that the *Yearbook* appears in print.

Of the Jupiter probes, Pioneer 10 continues its flight out of the Solar System, while Pioneer 11 is still en route to rendezvous with Saturn. Further probes are being planned by the Americans, and no doubt by the Russians also.

Another object which attracted great attention during the period under review (August 1975 – August 1976) was Nova (V. 1500) Cygni, the brightest nova for many years. It was discovered on 29 August 1975 by Honda, in Japan, at a time when daylight made observations from Europe impossible. When darkness fell, it was independently found by at least fifty observers in Britain alone – and indeed it could hardly be overlooked, since it was of almost the second magnitude. Its rise had been of unparalleled rapidity, and there were even suggestions (soon, alas, dispelled) that it might be a supernova. On the following night it reached magnitude 1·8, but its fall was as dramatic as its upsurge; it had dropped to the fifth magnitude by 4 September, and was below naked-eye visibility by 7 September. The magnitude continued to decrease, though less quickly, and had fallen to 9 by the end of the year. Nova Cygni was found to be about 3000 light-years away. An extra point of interest was its intense red colour for some time during early September.

Quite different was the star to which attention was drawn by O. J. Eggen, Director of the Mount Stromlo and Siding Spring Observatories in Australia. The magnitude of the star is a mere 10·8, and there are no suggestions that it is variable; but its intrinsic luminosity is extremely low, and it is certainly very close. At one time there were vague suggestions that it might be even closer than Alpha Centauri. This is probably not the case, but no exact estimates have yet been made. Eggen's Star is in Sculptor, about 1·1 degrees south-west of Tau Sculptoris.

Needless to say, investigations into quasars, BL Lacertæ objects and other important objects have continued apace. Considerable attention has been paid to the centre of our own Galaxy; observations with very long baseline interferometer techniques have indicated that there is a very small, intense radio source near the true centre, with a diameter which can hardly be much more than one light-day. As yet the investigations are at an early stage, and it would be unwise to jump to any dramatic conclusions (such as the favourite theory that we may be studying either a 'superstar' or a Black Hole), but it is true that the existence of some unusual object in this region is a possibility. There may even be some sort of analogy with the galaxy NGC 1275, which is 400,000,000 light-years away, and which has an active nucleus containing several very small radio sources.

Work of this kind requires highly elaborate equipment, and by now 'invisible astronomy' has become of paramount importance. Yet sheer visual work cannot be superseded, and during 1975-6 the great new Russian reflector was brought into full commission; Soviet statements indicate that it is working well. Whether it will remain the world's largest optical telescope remains to be seen.

West's Comet, photographed on 9 March 1976 by Paul Doherty. The stellar magnitude of the comet at this time was 2·5.

PART THREE

Miscellaneous

Some Interesting Telescopic Variable Stars

Star	R.A. h	R.A. m	Dec. °	Dec. '	Mag. range	Period, days	Remarks
R Andromedæ	0	22	+38	18	6·1–14·9	409	
W Andromedæ	2	14	+44	4	6·7–14·5	397	
R Aquilæ	19	4	+ 8	9	5·7–12·0	300	
R Arietis	2	13	+24	50	7·5–13·7	189	
R Aurigæ	5	13	+53	32	6·7–13·7	459	
R Boötis	14	35	+26	57	6·7–12·8	223	
R Cassiopeiæ	23	56	+51	6	5·5–13·0	431	
T Cassiopeiæ	0	20	+55	31	7·3–12·4	445	
T Cephei	21	9	+68	17	5·4–11·0	390	
Omicron Ceti	2	17	− 3	12	2·0–10·1	331	Mira.
R Coronæ Borealis	15	46	+28	18	5·8–14·8	—	Irregular
W Coronæ Borealis	16	36	+37	55	7·8–14·3	238	
R Cygni	19	35	+50	5	6·5–14·2	426	
U Cygni	20	18	+47	44	6·7–11·4	465	
W Cygni	21	34	+45	9	5·0– 7·6	131	
SS Cygni	21	41	+43	21	8·2–12·1	—	Irregular.
Chi Cygni	19	49	+32	47	3·3–14·2	407	Near Eta.
R Draconis	16	32	+66	52	6·9–13·0	246	
R Geminorum	7	4	+22	47	6·0–14·0	370	
U Geminorum	7	52	+22	8	8·8–14·4	—	Irregular
S Herculis	16	50	+15	2	7·0–13·8	307	
U Herculis	16	23	+19	0	7·0–13·4	406	
R Hydræ	13	27	−23	1	4·0–10·0	386	
R Leonis	9	45	+11	40	5·4–10·5	313	Near 18, 19.
X Leonis	9	48	+12	7	12·0–15·1	—	Irregular (U Gem type).
R Leporis	4	57	−14	53	5·9–10·5	432	'Crimson star.'
R Lyncis	6	57	+55	24	7·2–14·0	379	
W Lyræ	18	13	+36	39	7·9–13·0	196	
HR Delphini	20	40	+18	58	3·6– ?	—	Nova, 1967.
LV Vulpeculæ	19	45	+27	2	4·8– ?	—	Nova, 1968.
U Orionis	5	53	+20	10	5·3–12·6	372	
R Pegasi	23	4	+10	16	7·1–13·8	378	
S Persei	2	19	+58	22	7·9–11·1	810	Semi-regular.
R Scuti	18	45	− 5	46	5·0– 8·4	144	
R Serpentis	15	48	+15	17	5·7–14·4	357	
SU Tauri	5	46	+19	3	9·2–16·0	—	Irregular (R CrB type).
R Ursæ Majoris	10	41	+69	2	6·7–13·4	302	
S Ursæ Majoris	12	42	+61	22	7·4–12·3	226	
T Ursæ Majoris	12	34	+59	46	6·6–13·4	257	
S Virginis	13	30	− 6	56	6·3–13·2	380	
R Vulpeculæ	21	2	+23	38	8·1–12·6	137	

Some Interesting Double Stars

The pairs listed below are well-known objects, and all the primaries are easily visible with the naked eye, so that right ascensions and declinations are not given. Most can be seen with a 3-inch refractor, and all with a 4-inch under good conditions, while quite a number can be separated with smaller telescopes, and a few (such as Alpha Capricorni) with the naked eye. Yet other pairs, such as Mizar-Alcor in Ursa Major and Theta Tauri in the Hyades, are regarded as too wide to to be regarded as bona-fide doubles!

Name	Magnitudes	Separation,"	Position angle, deg.	Remarks
Gamma Andromedæ	3·0, 5·0		060	Yellow, blue. B is again double (0"·4) but needs a larger telescope.
Zeta Aquarii	4·4, 4·6	2·6	291	Becoming more difficult.
Gamma Arietis	4·2, 4·4	8	000	Very easy.
Theta Aurigæ	2·7, 7·2	3	330	Stiff test for 3 in. OG
Delta Boötis	3·2, 7·4	105	079	Fixed.
Epsilon Boötis	3·0, 6·3	2·8	340	Yellow, blue. Fine pair.
Kappa Boötis	5·1, 7·2	13	237	Easy.
Zeta Cancri	5·6, 6·1	5·6	082	
Iota Cancri	4·4, 6·5	31	307	Easy. Yellow, blue.
Alpha Canum Venat.	3·2, 5·7	20	228	Yellowish, bluish. Easy
Alpha Capricorni	3·2, 4·2	376	291	Naked-eye pair. Alpha again double.
Eta Cassiopeiæ	3·7, 7·4	11	298	Creamy, bluish. Easy.
Beta Cephei	3·3, 8·0	14	250	
Delta Cephei	var, 7·5	41	192	Very easy.
Xi Cephei	4·7, 6·5	6	270	Reasonably easy.
Gamma Ceti	3·7, 6·2	3	300	Not too easy.
Zeta Coronæ Borealis	4·0, 4·9	6·3	304	
Delta Corvi	3·0, 8·5	24	212	
Beta Cygni	3·0, 5·3	35	055	Yellow, green. Glorious.
61 Cygni	5·3, 5·9	25	150	
Gamma Delphini	4·0, 5·0	10	265	Yellow, greenish. Easy.
Nu Draconis	4·6, 4·6	62	312	Naked-eye pair.
Alpha Geminorum	2·0, 2·8	2	151	Castor. Becoming easier.
Delta Geminorum	3·2, 8·2	6·5	120	
Alpha Herculis	var, 6·1	4·5	110	Red, green.
Delta Herculis	3·0, 7·5	11	208	Optical double.
Zeta Herculis	3·0, 6·5	1·4	300	Fine, rapid binary.
Gamma Leonis	2·6, 3·8	4·3	121	Binary; period 400 years
Alpha Lyræ	0·0, 10·0	60	180	Vega. Optical; B faint.
Epsilon Lyræ	4·6, 6·3	3	005	Quadruple. Both pairs separable in 3 in. OG
	4·9, 5·2	2·3	111	
Zeta Lyræ	4·2, 5·5	44	150	Fixed. Easy double.
Beta Orionis	0·1, 6·7	9·5	205	Rigel. Can be split with 3 in.
Iota Orionis	3·2, 7·3	11	140	

SOME INTERESTING CLUSTERS AND NEBULÆ

Name	Magnitudes	Separation,"	Position angle, deg.	Remarks
Theta Orionis	6·0, 7·0 7·5, 8·0			The famous Trapezium in M.42
Sigma Orionis	4·0, 7·0 7·5, 10·0	11·1 12·9	236 085	Quadruple. D is rather faint in small apertures.
Zeta Orionis	1·9, 5·0	3	160	
Eta Persei	4·0, 8·5	8·5	300	Yellow, bluish.
Alpha Piscium	4·3, 5·3	1·9	291	
Alpha Scorpii	0·9, 6·8	3	275	Antares, Red. green.
Nu Scorpii	4·2, 6·5	42	336	
Theta Serpentis	4·1, 4·1	23	103	Very easy.
Alpha Tauri	0·8, 11·2	130	032	Aldebaran. Wide, but B is very faint in small telescopes.
Zeta Ursæ Majoris	2·3, 4·2	14·5	150	Mizar. Very easy. Naked eye pair with Alcor.
Alpha Ursæ Minoris	2·0, 9·0	18·3	217	Polaris. Can be seen with 3 in.
Gamma Virginis	3·6, 3·7	4·8	305	Binary; period 180 yrs. Closing.
Theta Virginis	4·0, 9·0	7	340	Not too easy.

Some Interesting Clusters and Nebulæ

Object	R.A.		Dec.		Remarks
	h	m	°	'	
M.31 Andromedæ	00	40·7	+41	05	Great Galaxy, visible to naked eye.
H.VIII 78 Cassiopeiæ	00	41·3	+61	36	Fine cluster, between Gamma and Kappa Cassiopeiæ.
M.33 Trianguli	01	31·8	+30	28	Spiral. Difficult with small apertures.
H.VI 33–4 Persei	02	18·3	+56	59	Double cluster; Sword-handle.
M.1 Tauri	05	32·3	+22	00	Crab Nebula, near Zeta Tauri.
M.42 Orionis	05	33·4	−05	24	Great Nebula. Contains the famous Trapezium, Theta Orionis.
M.35 Geminorum	06	06·5	+24	21	Open cluster near Eta Geminorum.
H.VII 2 Monocerotis	06	30·7	+04	53	Open cluster, just visible to naked eye.
M.41 Canis Majoris	06	45·5	−20	42	Open cluster, just visible to naked eye.
M.44 Cancri	08	38	+20	07	Præsepe. Open cluster near Delta Cancri. Visible to naked eye.
M.97 Ursæ Majoris	11	12·6	+55	13	Owl Nebula, diameter 3'. Planetary.
M.3 Canum Venaticorum	13	40·6	+28	34	Bright globular.
M.80 Scorpionis	16	14·9	−22	53	Globular, between Antares and Beta Scorpionis.
M.4 Scorpionis	16	21·5	−26	26	Open cluster close to Antares.
M.13 Herculis	16	40	+36	31	Globular. Just visible to naked eye.
M.92 Herculis	17	16·1	+43	11	Globular. Between Iota and Eta Herculis.
M.7 Scorpionis	17	51·6	−34	48	Fine open cluster. Very low in England.
M.23 Sagittarii	17	54·8	−19	01	Open cluster nearly 50' in diameter.
H.VI 37 Draconis	17	58·6	+66	38	Bright Planetary.
M.8 Sagittarii	18	01·4	−24	23	Lagoon Nebula. Gaseous. Just visible with naked eye.
NGC 6572 Ophiuchi	18	10·9	+06	50	Bright planetary, between Beta Ophiuchi and Zeta Aquilæ.
M.17 Sagittarii	18	18·8	−16	12	Omega Nebula. Gaseous. Large and bright.
M.11 Scuti	18	49·0	−06	19	Wild Duck. Bright open cluster.
M.57 Lyræ	18	52·6	+32	59	Ring Nebula. Brightest of planetaries.
M.27 Vulpeculæ	19	58·1	+22	37	Dumb-bell Nebula, near Gamma Sagittæ.
H.IV 1 Aquarii	21	02·1	−11	31	Bright planetary near Nu Aquarii.
M.15 Pegasi	21	28·3	+12	01	Bright globular, near Epsilon Pegasi.
M.39 Cygni	21	31·0	+48	17	Open cluster between Deneb and Alpha Lacertæ. Well seen with low powers.

Some Recent Books

Infra-red: The New Astronomy, by David A. Allen. Keith Reid Ltd (distributed by David & Charles Ltd), 1976. The first comprehensive account of this vitally important new branch of astronomy, written in a way which will be intelligible to the layman as well of great value to the expert.

Evolution of Stars and Galaxies, by Walter Baade. MIT Press, 1975. A new edition of this classic book, edited by Cecilia Payne-Gaposchkin. The original lectures by the late Dr Baade have been brought up to date.

Guide to the Moon, by Patrick Moore. Lutterworth Press, 1976. The new edition of this book, completely re-written after the Apollo landings and with a revised lunar map for observers.

Eyes on the Universe, by Isaac Asimov. Andre Deutsch, 1976. A history of the development of the telescope, written at popular level, but with so much information as to be useful also to the scholar.

The Universe: Its Beginning and End, by Lloyd Motz. Scribner, 1976. A comprehensive introduction, for the informed layman.

Our Contributors

ERNEST ROY TURNER is a retired banker. As a resident of Burstow since 1929 he has always taken an active part in local affairs. Educated at Sir Walter St John's, London, he was elected to the Burstow Parish Council in 1949 and served for 19 years, six as Chairman. He served in the Territorial Army, was commissioned in the R.A.F.V.R. in 1937, and was invalided out after war service in 1946. He is a member of the Surrey Local History Council and the Horley Local History Association, and is now starting work upon a full history of Burstow.

H. G. MILES, B.E.M., B.Sc., is a lecturer at the Lanchester Polytechnic, Coventry. He was President of the British Astronomical Association, 1974-6, and is Director of its Artificial Satellite Section. He also specializes in meteor and meteorite research. He is the author of many scientific contributions, both technical and popular, and will be well known to all readers of past *Yearbooks*.

DR GARRY HUNT, of the Meteorological Office, Bracknell, is one of the world's leading experts in planetary research, and is closely concerned with the current space-probe projects, so that he makes frequent visits to the American centres. In addition to his many technical contributions he is also well known for his television broadcasts.

DR PETER VAN DE KAMP is Director of the Sproul Observatory, Swarthmore College, Pennsylvania. His contributions to astronomy in many aspects have been outstanding, and it was he who first established the existence of planets moving round other stars.

DR MARTIN COHEN is a Cambridge graduate. He concentrates upon infra-red astronomy, and is now continuing his research in the United States; he was formerly on the staff of the Royal Greenwich Observatory, and has made many original contributions.

DR DAVID A. ALLEN took his B.A. at Cambridge University in 1967, and his Ph.D. in 1971. After some years at the Royal Greenwich Observatory, he is now continuing his research in Australia. His recent work has been concentrated upon infra-red astronomy; telescopes used include the Palomar 200 in. and the Mount Wilson 100 in., as well as the 98-in. Isaac Newton reflector.